The MATH EXPLORER

Games and Activities for Middle School Youth Groups

An Exploratorium Book

Pat Murphy, Lori Lambertson, Pearl Tesler, and the Exploratorium

WITHDRAWN

 expl◯ratorium

 **Key Curriculum Press**
Innovators in Mathematics Education

This material is based upon work supported by the National Science Foundation under Grant No. ESI-9902047. Any opinions, findings, and conclusions or recommendations expressed in this publication are those of the author(s) and do not necessarily reflect the views of the National Science Foundation.

BE CAREFUL! The activities and projects in this book were designed with safety and success in mind. But even the simplest activity or the most common materials can be harmful when mishandled or misused. Follow safety instructions and adapt procedures to the abilities of your group whenever you're exploring or experimenting.

Exploratorium
3601 Lyon Street
San Francisco, CA 94123
www.exploratorium.edu

Key Curriculum Press
1150 65th Street
Emeryville, CA 94608
510-595-7000
editorial@keypress.com
http://www.keypress.com
Printed in the United States of America

10 9 8 7 6 5 4 3 2   07 06 05 04

ISBN 1-55953-540-7

Project Editor: Joan Lewis
Editorial Assistants: Erin Gray, Susan Minarcin
Consulting Editor: Mali Apple
Mathematics Consultants: Larry Copes, Paul Doherty, Thomas Humphrey
Editorial Production Manager: Deborah Cogan
Production Editor: Jacqueline Gamble
Copyeditor: Tom Briggs
Production Director: Diana Jean Parks
Production Coordinator, Compositor: Mike Hurtik
Text Designer: Carolyn Deacy
Illustrator: Jason Gorski
Art and Design Coordinator: Caroline Ayres
Cover Designer: John Nedwidek
Prepress and Printer: Versa Press, Inc.

Executive Editor: Casey FitzSimons
Publisher: Steven Rasmussen

Exploratorium
Manager of Learning Tools: Kurt Feichtmeir
Assistant Editor: Laura Jacoby
Administrative Support: Megan Bury
Executive Associate Director: Rob Semper
Director of the Exploratorium: Goéry Delacôte

Contents

Introduction

The Exploratorium—San Francisco's hands-on museum of science, art, and human perception—has more than 650 exhibits, and all of them run on curiosity. At the Exploratorium, people learn by experimenting, by asking questions and finding their own answers, and by playfully exploring the world around them.

The Exploratorium is well known for innovative science exhibits and science teaching. In this book, we apply the hands-on teaching methods that work so well in science to teaching and learning about mathematics. This book offers 24 fun, creative, hands-on activities in mathematics for groups of middle schoolers and their leaders.

The activities in this book will get learners playing, thinking, building, and experimenting with mathematics and how it works. Flip through the pages, and you'll find games to play, structures to build, toys to create, and puzzles to solve. At the heart of each project or activity is a mathematical skill or process. These activities allow participants to experiment with mathematics without the pressure of timed tests and without a high premium on the amount of knowledge they bring to the activity.

Many people have real anxieties about their ability to do mathematics. How often have you heard someone say, "I've never been any good at math," or even "I just hate math," and seen others nod in agreement? Fear of math influences people's choices, limiting what they can do in their careers and even in their everyday lives.

We created this book in part to alleviate that fear. We designed the activities for middle schoolers because giving up on mathematics at that level closes off many academic opportunities and career paths. We think that playing with mathematical ideas in games and activities outside school can help middle schoolers become more comfortable with the same material when they encounter it in school.

If you feel challenged by mathematics yourself, don't worry. You don't have to be "good in math" to help your group make beautiful boxes from old greeting cards or launch rockets made from recycled paper. For those leaders who want to delve into the mathematical implications of these activities, we have included streamlined explanations of the mathematics in the "Tips for Leaders" sections.

When we began working on this book, we talked with youth development professionals about what they needed for their programs. We heard that many were looking for mathematics activities that would appeal not only to the young people participating in their programs but also to parents, teachers, program funders, and administrators.

To appeal to young people, activities have to be engaging and fun. But at the same time, the activities need to meet local, state, and national standards for mathematics teaching in middle school. We believe we've created activities that do so.

All the activities in this book have been tested with groups of young people. We've included only those they liked best. The resulting collection of games, puzzles, experiments, and projects provides opportunities for practicing a variety of mathematics skills—from problem solving and graphing to fractions and ratios—*and* for having fun.

Mathematics is a creative, exciting subject. Like science, it gives people a perspective on how the world works and a foundation for lifelong learning. The ability to use mathematics in everyday life and in the workplace has never been more important. We hope the ideas in these activities will give learners and leaders alike new ways of thinking about mathematics.

Read This First!

If you're like everyone at the Exploratorium, you'll want to skip all the introductory stuff and get right to the fun. But reading this section first can save you a lot of time and trouble later.

Important Things to Know Before You Begin

- You *don't* have to do the activities in any special order.
- You *don't* have to "know" the mathematics.
- You *don't* have to make photocopies.

Pages for You, Pages for Your Group

Each activity in this book has three simple parts that include information for you and your group.

What Is It?

The first page of each activity tells you all about that activity. The "Planning Chart" tells how much time it will take. A "Preparation and Materials" section lists the things you'll need. You'll also find ideas for getting started and a description of the kinds of mathematics skills involved in the activity.

Explorer's Notebook

The next few pages, titled "Explorer's Notebook," are easy-to-follow, step-by-step instructions for learners. You can copy them for your group if you want, but you don't have to. If you prefer, you can read the material and then demonstrate each step. This is a good option if your group struggles with reading in English or if you don't have access to a photocopier.

Some of these pages are data sheets or game boards, designed to be copied for use in the activity. If you don't have access to a photocopier, however, you can have members of your group copy them by hand, making their own game boards and data sheets. If you want to use the games over an extended period, you may want to copy the game boards on card stock and laminate them.

Tips for Leaders

The last pages of each activity contain "Tips for Leaders." Here you will find suggestions on how to use the activities, advice on how to avoid potential problems, ideas on how to extend the activities, and additional information from historical background to science connections.

"Where's the Math?" gives you simple, straightforward explanations of the mathematics involved in each activity. If you want to check a particular word in your mathematics vocabulary, there's also a handy glossary at the back of the book.

The Teaching Connection

Despite an emphasis on creativity and fun, the activities in this book represent serious learning. They are designed to reinforce mathematics skills and meet the National Council of Teachers of Mathematics (NCTM) *Principles and Standards for School Mathematics*. For a detailed look at how each activity can help support mathematics learning, check out the chart of NCTM Content and Process Standards at the back of the book.

The Right Stuff: A Note About Materials

We've designed these activities so that you can do them on a shoestring budget. If you want to purchase materials you don't have on hand, you'll find everything readily available in grocery and variety stores, home and office supply outlets, art supply stores, gift shops, toy stores, and hardware stores. Also, please visit *The Math Explorer* Web site at www.keypress.com/ME for more information about *The Math Explorer* kit.

Getting Started

You don't have to do these activities in any special order. Each was designed to stand on its own, so you can start just about anywhere. Pick something that looks like fun, or check out "Which Activities Should I Do?" on page xi for suggestions on specific activities that might be best for particular situations.

Which Activities Should I Do?

If you're having trouble figuring out where to start, here are a few suggestions that might help.

If You Have No Time to Prepare

Several of the activities in this book require minimal preparation. If you don't have any time to get ready, we suggest you try one of these:

- Boxed In!
- Oddball
- Pig
- Hopping Hundred
- Tic-Tac-Toe Times
- Fantastic Four

If Your Group Needs Practice with the Multiplication Table

If members of your group struggle with the multiplication table, these activities may benefit them:

- Eratosthenes' Sieve
- Hopping Hundred
- Tic-Tac-Toe Times
- Boxed In!

If Individuals Have Extra Time

If people need something to do on their own, here are activities that one or two people can do without disturbing others:

- Madagascar Solitaire
- Magic Grid
- Mind Reader

If People Want to Make Something to Take Home

A number of activities involve making things to take home.

- Incredible Shrinking Shapes
- Greeting Card Boxes
- Jacob's Ladder
- Colossal Cartoons
- Tetrahedral Kites
- Paper Engineering

If Your Group Wants to Experiment to Figure Things Out

These activities combine mathematics and science, so your group can experiment to improve their mathematics skills.

- Flying Things
- Height Sight
- Stomp Rocket!

Playing Games

Playing games may not seem to have much to do with math—but it can. All the games in this section encourage people to think ahead, analyzing possible moves and their outcomes. Though anyone can play these games and have a good time, the player who thinks about the game and plans ahead is more likely to win. This sort of strategic planning is part of thinking mathematically. These games also encourage people to work on basic mathematical skills.

Activity 1 ## Boxed In!

This entertaining, easy-to-learn game gives players multiplication practice and increases their understanding of area, an important concept in middle school geometry.

Activity 2 ## Oddball

This simple game encourages players to think ahead. With a little analysis, they can figure out how to win every time! Variations on the game will challenge any player.

Activity 3 ## Pig

This dice-rolling game gives players a chance to practice doing addition in their heads. Pig *also offers them an opportunity to explore the role of probability in determining the chances of rolling a certain number.*

Activity 4 ## Madagascar Solitaire

This ancient solitaire game encourages players to think about strategy and symmetry—and also teaches skills related to graphing.

Activity 5 ## Fantastic Four

This game gives players a chance to practice addition, subtraction, multiplication, and division. More important, it invites them to work on their problem solving skills, figuring out creative solutions to tricky problems.

Boxed In!

In this fun game, players compete to see who can fit the most rectangles into a grid. The game gives players multiplication practice and increases their understanding of area, an important concept in middle school geometry.

Preparation and Materials

You can have your whole group play *Boxed In!* together, or you can provide instructions for small groups of two or more to play independently.

To play as a group, you will need:

- a pair of dice
- fine-tipped colored markers or pens (at least 1 per person)
- copies of *Boxed In! Game Grids* (Each copy has 3 grids, enough to play 3 games; you will need at least 1 per person.)

To play in small groups, you will also need:

- a pair of dice for each group (Players can make *Paper Dice* by following the instructions on page 195.)

Using This Activity

Several group leaders who tested this game suggested having extra *Boxed In! Game Grids* on hand. Their groups went through three games very quickly.

Group leaders found that people enjoyed playing *Boxed In!* for about half an hour. If you have more time, combine this game with *Pig* (page 14), another game that uses dice.

More tips for how to use *Boxed In!* start on page 5.

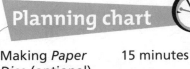

Planning chart

Making *Paper Dice* (optional)	15 minutes
Playing *Boxed In!* (three games)	20–30 minutes

Playing Boxed In!

In this game, a roll of the dice gives you the size of a rectangle. The player who can fit the most rectangles on the grid wins.

What Do I Need?

◇ a pair of dice

◇ a colored marker or pen

◇ a copy of *Boxed In! Game Grids*

Here's How to Play

1 A player rolls the dice and reads the numbers that come up.

2 On the grid, each player draws a rectangle with measurements that match the numbers on the dice. Suppose the dice come up 3 and 5. Draw a rectangle that is 3 squares high and 5 squares wide—or 5 squares high and 3 squares wide. Inside this rectangle, write its dimensions and area.

3 Another player rolls the dice. Using the numbers rolled, draw another rectangle on the grid. This rectangle can't overlap the first rectangle. Write the rectangle's dimensions and area inside the rectangle.

4 Take turns rolling the dice. Every time the dice are rolled, draw another rectangle and write the length, width, and area inside it. Once a rectangle is drawn on the grid, you can't move it.

5 After a while, your grid will get crowded. When you can't fit in another rectangle, you are out of the game. The last player who can still fit a rectangle on his or her grid wins the game.

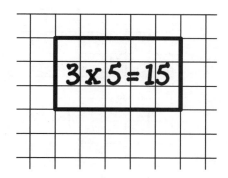

You can draw the 3-by-5 rectangle on your grid like this . . .

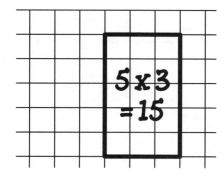

. . . or like this.

Boxed In! Game Grids

Game 1

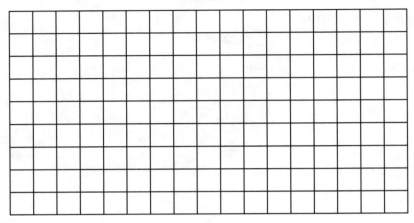

Game 2

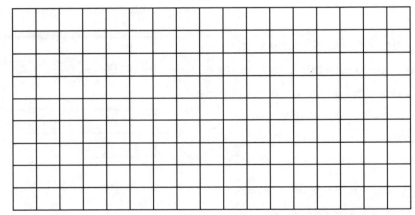

Game 3

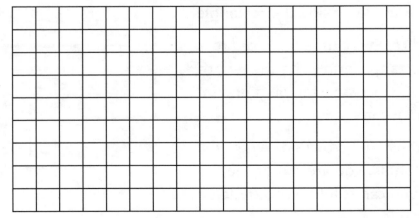

Playing *Boxed In!*

You can play *Boxed In!* with any size group. Use the basic game to encourage practice of multiplication skills. Add even more math practice—and more fun—by introducing the variation described on page 6.

Playing the Game

Even if you want your group to split up into small groups to play, it's good to begin by playing one game with the entire group. That way, you make sure everyone understands how to play.

Explain how the game works to the group. As the leader, you can either roll the dice yourself or have players take turns rolling. If you have a chalkboard, write down the numbers from each roll. Anyone who falls behind can use that list to make sure they've drawn all the proper rectangles.

Once a rectangle is drawn on a grid, a player is not allowed to move it. That's why we suggest having players use pens or markers, not pencils. One leader who tested this game had people use a different color for each roll of the dice. The different colors made it easy for people to check their work at the end.

The last player who can fit a rectangle on his or her grid wins. There will often be multiple winners. When the dice come up with a large rectangle, many players will be eliminated at once.

When you have finished a game, you might have everyone count all the squares that remain. All players should have the same number of squares left over.

Encouraging Math Practice

The space enclosed by a rectangle is the *area* of that rectangle. When the rectangle is drawn on a grid, the area is the number of squares inside the rectangle. The area of a rectangle is calculated by multiplying the rectangle's dimensions. The dimensions of a rectangle are length and width. So, multiply the rectangle's length by its width. For example, the area of a 3-by-5 rectangle is calculated like this:

$$3 \times 5 = 15$$

If you want members of your group to practice multiplication, ask them to call out the area of the rectangle after each roll of the dice.

To add even more math practice to the game, have players keep track of the total area that is used up as the game progresses. The game grid measures 17-by-9, so it has an area of 153 squares. Start by writing the beginning area of 153 squares on the board. After each roll, calculate the area of the new rectangle and subtract it from the total area remaining.

This is a great way to check whether players are making mistakes. If a player claims to fit in a rectangle that's larger than the remaining area, something is very fishy!

Strategy in *Boxed In!*

Your group will quickly realize that the way to fit the most rectangles on the grid is to conserve space. By squeezing the rectangles as close together as possible or by running thin ones along a side of the grid, they will leave more space for new rectangles.

You might talk about how playing *Boxed In!* is a little like putting boxes in a closet. To fit the most boxes, you pack them close together in an organized way.

Variation on *Boxed In!*

After your group has played *Boxed In!* a few times, you might introduce this variation, invented by Naomie Wright at San Francisco's Omni/Excelsior Neighborhood Beacon Center.

Once players calculate the area of a rectangle described by a roll of the dice, tell them they can now draw *any* rectangle with that area on their grids. If the dice come up 3 and 4, for example, the area is 12. Players can draw a rectangle measuring 2-by-6, or 1-by-12, or 3-by-4.

Where's the Math?

Boxed In! gives players an opportunity to practice multiplication skills and to think strategically.

This game also reviews how to calculate the area of a rectangle given its dimensions. If you try the variation on the game, players will also learn that rectangles of different shapes can have the same area.

When you tell your group to calculate the area of a rectangle by multiplying the length by the width, some people may feel they are counting the corner squares twice. You can show them that this isn't so. Draw a 4-by-6 rectangle on a grid. Calculate the area by multiplying 4 and 6. You get 24. Now count the squares inside the rectangle. Once again, you will get 24!

The Math Explorer
Published by Key Curriculum Press / © 2003 Exploratorium

...A C T I V I T Y 2

Oddball

This game encourages people to think strategically and analyze the consequences of each move. Believe it or not, *Oddball* can be won or lost in the very first move! After players figure out how to win the game, they can move on to more challenging variations.

Preparation and Materials

For each pair of players, you will need:

*   10 identical objects to use as playing pieces (pennies, beans, paper clips, or other small objects)

*   copies of *Playing Oddball* and *Variations on Oddball* (optional)

To explore winning strategies, you will also need a chalkboard where you can copy the *Analyzing Oddball* chart or a copy of *Analyzing Oddball* for each player.

Using This Activity

You can play *Oddball* with a group, or it can be played independently by groups of two, making it a great activity for people who finish other assignments early.

Players may quickly realize that there is a strategy to winning *Oddball*. Going through *Analyzing Oddball* with your group will ensure that everyone understands how to win. Even people who figure out *Oddball* quickly will be challenged by *Variations on Oddball*, so it's well worth encouraging your group to give these other games a try.

More tips for how to use *Oddball* start on page 11.

Planning chart	
Playing *Oddball*	20 minutes
Analyzing *Oddball*	20 minutes
Variations on *Oddball*	open-ended

Playing Oddball

In this game, no one wants to be left with the last playing piece—the "oddball." Can you figure out how to avoid getting stuck with the last piece?

What Do I Need?

◇ 10 identical objects to use as playing pieces (pennies, paper clips, beans, or other small objects)

◇ a partner

Here's How to Play

1 Put 10 playing pieces in a pile on the table.

2 One player starts by taking either one or two pieces from the pile.

3 The other player takes either one or two pieces from the pile.

4 Players take turns removing pieces from the pile. On each turn, you can take either one or two pieces, but you have to take at least one.

5 The player who is forced to take the last piece loses the game.

Believe it or not, you can win or lose this game in the very first move. Can you figure out how? Play *Oddball* a few times, and then analyze the game to learn more.

Analyzing Oddball

In playing *Oddball,* you may notice that certain moves always lead to winning while other moves don't. To figure out a strategy that lets you win every time, work with your leader to fill out this chart.

Pieces Left at the Start of Your Turn	Win or Lose?
1	
2	
3	
4	
5	
6	
7	
8	
9	
10	

Variations on Oddball

Even after you've mastered *Oddball,* you'll find these variations on the game to be challenging.

Going in Circles

This game is a geometric variation on *Oddball.* To play it, you will need 10 identical playing pieces.

Here's How to Play

1 Arrange the 10 playing pieces in a circle.

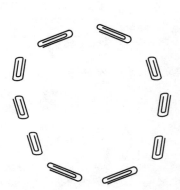

2 Players take turns removing one or two pieces from the circle, just as in *Oddball.* But there's one more rule: a player can take two pieces only if they are right next to each other.

3 The player to take the last piece loses.

Nim

Nim is the classic take-away game. To play it, you'll need 15 identical playing pieces.

Here's How to Play

1 Arrange the 15 playing pieces in three groups:
   ◇ one group of 3
   ◇ one group of 5
   ◇ one group of 7

2 Players take turns removing pieces from these groups. During a turn, a player can take as many pieces as he or she wants, but all must be taken from the same group.

3 The player who takes the last piece wins.

Playing and Analyzing *Oddball* with a Group

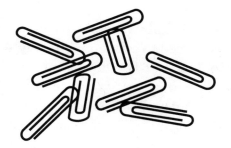

Math is more than simply manipulating numbers. It's a way of thinking about the world and a method for solving problems. This game gives people a chance to practice thinking strategically—planning ahead and figuring out the consequences of their moves. Analyzing the game teaches problem solving skills.

Playing *Oddball*

Start by reading "Here's How to Play" (page 8) aloud. Then have players pair up and play a few games. Students new to English may wonder about the word *oddball*. It refers to someone who does not follow the usual style.

After everyone has played a few games, discuss the strategy behind the game. Ask if anyone discovered a technique for winning. Are there certain situations that *always* lead to a win or a loss?

Noticing Patterns While Playing

As people play *Oddball*, they will probably start to notice certain patterns. For instance, they might realize that a player faced with three pieces at the start of a turn can always win by taking two pieces.

If your group notices this, encourage members to look for an earlier point in the game when a player can't possibly win. Chances are, they'll notice that a player faced with four pieces at the start of a turn will lose when playing a savvy opponent. If the player

takes one piece, the opponent is left with three. If the player takes two pieces, the opponent is left with two. Either way, the opponent can easily win.

People may figure out winning strategies on their own. If they don't, you might want to lead them through an analysis of the game.

Analyzing *Oddball*

The *Analyzing Oddball* chart lets people keep track of what they figure out about the game. With your help, your group can fill out the chart to devise a strategy that makes it possible to win every time.

If you have a chalkboard, you can copy the *Analyzing Oddball* chart onto the board and keep track of the results there. Or you can have each person fill out a copy of *Analyzing Oddball*.

We suggest you walk your group through the process of filling out the chart by asking such questions as these. (The answers are in parentheses.)

- Suppose it's your turn and there's one piece left. (You lose. Write "You lose" next to "1" on the chart.)

- What if it's your turn and there are two pieces left? (You take one, leaving your opponent with the last piece. And you win. Write "Take 1 to win" next to "2" on the chart.)

- What if it's your turn and there are three

pieces left? Can you win? (To be sure to win, you need to take enough away so that your opponent has a "losing number." If you take two pieces, you leave one for your opponent, and you win. Write "Take 2 to win" next to "3" on the chart.)

- What if it's your turn and there are four pieces left? You want to leave your opponent with a losing number. Can you do it? (No. You can take one piece and give your opponent three, a winning number. Or you can take two pieces and give your opponent two, another winning number. So, if you are faced with four pieces, you will lose if your opponent is thinking ahead. Write "You lose" next to "4" on the chart.)

- What if it's your turn and there are five pieces left? (Take one and leave your opponent with four, a losing number. If your group gets confused, remind them that they want to leave their opponent with a number that says "You lose" beside it.)

- What if it's your turn and there are six pieces left? (Take two and leave your opponent with four, a losing number.)

- What if it's your turn and there are seven pieces left? (You can't take away enough pieces to leave your opponent with a losing number. If you can't take away enough pieces to reach a losing number, then you are on a losing number. Write "You lose" next to "7.")

- What if it's your turn and there are eight pieces left? (Take one and leave your opponent with seven, a losing number.)

- What if it's your turn and there are nine

Pieces Left at the Start of Your Turn	Win or Lose?
1	You lose
2	Take 1 to win
3	Take 2 to win
4	You lose
5	Take 1 to win
6	Take 2 to win
7	You lose
8	Take 1 to win
9	Take 2 to win
10	You lose

pieces left? (Take two and leave your opponent with seven, a losing number.)

- What if you have the first turn and there are ten pieces? What do you do? (You can't take away enough pieces to leave your opponent with a losing number, which means that you are on a losing number. Write "You lose" next to "10." Because ten is a losing number, you don't want to go first when playing *Oddball*.)

The Math Explorer
Published by Key Curriculum Press / © 2003 Exploratorium

Ask group members if they see a pattern in the chart. Can they use the pattern to guess the result for 11, 12, or 13 pieces?

What's Next?

After analyzing the game, everyone should see that a player can be sure to win if he or she goes second. If the opponent goes first, he or she will take either one or two pieces, leaving either eight or nine pieces. Because both 8 and 9 are winning numbers, the player who goes second can always win—if he or she plays wisely.

You might ask what would happen if you changed the rules to allow players to take one, two, or three pieces. Have them play a game and analyze how the strategy for winning changes. With this change in the rules, is it still a good idea to go second?

Once the group analyzes *Oddball*, it won't be much fun to play. Have them play *Going in Circles* or *Nim*, challenging variations that are much more difficult to analyze.

Where's the Math?

An important part of math is being able to recognize patterns, especially patterns in series of things. Finding patterns can help you win games (like *Oddball*), solve puzzles (like *Madagascar Solitaire*; see page 25), and predict what will happen in certain situations (like *Flying Things*; see page 174). Analyzing complex patterns—like the movements of the planets in the solar system—is one use of math.

Pig

Playing *Pig* is a fun way for people to practice doing addition in their heads. This "mental math" skill is something many people need practice with.

My score was 75 . . . I rolled 8, so my new score is 8 plus 75. 8 + 75 = 83. I think I'll roll again!

Preparation and Materials

For each group of players, you will need:

- a pair of dice (Players can make *Paper Dice* using the template on page 197.)

- a copy of *Playing Pig*

To explore the concept of probability with your group, you'll also need:

- *Experimenting with Pig Probabilities* and/or *Exploring Pig Probabilities* (1 per player)

- for *Experimenting with Pig Probabilities*, a copy of page 16 that you can write on

- for *Exploring Pig Probabilities*, crayons, pens, or markers of at least 4 colors for each player (Having 11 colors is ideal.)

Using This Activity

Tips for using *Pig* start on page 19. If people have a hard time reaching 100 and get frustrated, change the winning score to 50.

You may want to have a *Pig* tournament. People play in pairs, and then the winners of each game play each other until you have a champion.

Work with your group to figure out strategies for winning *Pig* using the *Pig Possibilities* chart on page 17. This introduces probability, a topic covered in middle school math. To prepare, read pages 19–24.

Planning chart

Making *Paper Dice* (optional)	15 minutes
Playing *Pig*	20–30 minutes
Experimenting with *Pig Probabilities*	15 minutes
Exploring *Pig Probabilities*	15 minutes

Playing Pig

In this dice-rolling game, you win by getting the score closest to 100 points—*without going over 100*. If you get "piggy" by going over 100, you lose!

My score was 75 . . . I rolled 8, so my new score is 8 plus 75. 8 + 75 = 83. I think I'll roll again!

What Do I Need?

◊ a pair of dice

◊ a partner

Here's How to Play

1 To start, each player rolls one of the dice. The player who rolls the higher number goes first.

2 When it's your turn, roll both dice. Add the two numbers together. That's your score.

3 Roll again, and add that total to your previous score. Add the numbers in your head, and keep track of your score out loud. Don't write it down. Part of the challenge of this game is doing all the math in your head.

4 You can roll as many times as you want on a turn *unless you roll a 1*.
   ◊ If one of the dice comes up 1, you get zero points for that turn, and it's the other player's turn.
   ◊ If both dice come up 1s, that's "snake eyes." You lose all your points, your score goes to zero, and it's the other player's turn.

5 You can decide to stop rolling and pass the dice to the other player at any time. Don't forget—if your score goes over 100, you lose! If you get close to 100, you might want to let the other player have a turn. After your opponent takes a turn, you can decide whether you want to roll again.

6 The winner is the player whose final score is nearer to 100 (or exactly 100) *without going over*.

Experimenting with Pig Probabilities

When you roll a pair of dice, what number is most likely to come up? The answer to this question is the secret to winning *Pig.* You can experiment to figure out the answer.

Step 1 You and your partner will roll the dice 20 times and keep track of the numbers you roll. One person rolls the dice. The other person adds the numbers on the dice together and makes a mark beside that number on the chart below.

Sum of Two Dice	Marks to Show How Many Times You Rolled That Sum	Total
2		
3		
4		
5		
6		
7		
8		
9		
10		
11		
12		

Step 2 Add up how many times you rolled each number. Record this figure under "Total" in the chart.

Step 3 Work with your leader and the rest of your group to make a graph of everybody's results. Based on this graph, what can you say about how likely you are to roll a certain number? What can you say about the dice based on this graph?

The Math Explorer
Published by Key Curriculum Press / © 2003 Exploratorium

Exploring Pig Probabilities

Understanding the probability that a certain number will come up can help you figure out strategies for winning *Pig*.

What Do I Do?

Step 1 This chart shows all the ways that the numbers on two dice can be added. Each sum on the chart shows one possible outcome—one thing that could happen when you roll the dice. How many different outcomes are there when you roll two dice?

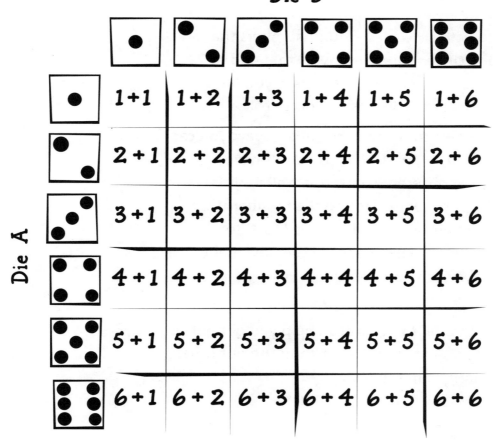

Pig Possibilities

Step 2 When you roll two dice, do some numbers come up more often than others? Which numbers come up most often?

To find the answers to these questions in the *Pig Possibilities* chart, you'll need some crayons, colored pens, or markers.

Use one color to circle all the different ways the numbers on two dice can add up to 7. You could roll 1 on Die A and 6 on Die B to get 7. Or you could roll 3 on Die A and 4 on Die B to get 7. How many other ways can you get 7?

Step 3 Now use a different color to circle all the ways you can get 2.

You'll end up with six circles around numbers that add to 7 and only one circle around numbers that add to 2. This means that each time you roll the dice, you have six times as many chances to roll 7 as you do to roll 2.

Step 4 Use different colors to circle all the ways you can roll 3, 4, 5, 6, 8, 9, 10, 11, and 12. If you don't have 11 different colors, use different shapes, as well as different colors. You could put squares around the numbers that add to 3 and circles around those that add to 4, for example. The idea is to make it easy to see how many ways you can make each number.

Which number are you most likely to roll?

Step 5 The *Pig Possibilities* chart can help you get better at playing *Pig*. Work with your leader to figure out other things you can learn from this chart.

The Math Explorer
Published by Key Curriculum Press / © 2003 Exploratorium

Using *Pig* to Explore Probability

You can use the game of *Pig* to give people a chance to practice mental addition. They'll do that every time they play. You can also use this game to talk about probability, a concept that is introduced in middle school math classes.

Some leaders have expressed concern that playing *Pig* with dice may encourage gambling. At the Exploratorium, we feel that an understanding of probability will have the opposite effect by showing people how often they will lose in "games of chance."

Playing *Pig*

Start by reading through "Here's How to Play" (page 15) with your group. Your group can divide into pairs, or members can play in threes, with the third person working through the math on paper to catch mistakes. This person can play the winner of each round.

What Is Probability?

Probability is a way to measure how likely something is to happen. When you flip an ordinary coin, you have an equal chance of getting heads or tails. That is, your probability of getting heads is the same as your probability of getting tails.

If you flip a coin a hundred times, it's likely that you'll get about as many heads as tails. But probability doesn't guarantee equal numbers of heads and tails. Probability doesn't tell you what *will* happen. It merely tells you *how likely* something is to happen.

Where's the Math?

The most obvious math in *Pig* is the simple arithmetic—players must practice adding in their heads. But playing *Pig* also encourages people to think strategically—to look ahead and figure out how the next roll will help or hurt their chances of winning. Trying to predict the results of a roll leads to an understanding of probability.

When rolling a single die, what's the probability that a certain number will come up? A die has six sides, with a different number on each. In rolling one die, each number has an equal chance of coming up.

Many people who don't often play with dice think that the same is true for two dice, but it isn't. You and your group can discover this for yourselves with *Experimenting with Pig Probabilities* (page 16) and *Exploring Pig Probabilities* (page 17).

Experimenting with Pig Probabilities gives your group a chance to measure probabilities by experimenting and then graphing the results. *Exploring Pig Probabilities* lets people see what is possible with each roll of the dice—and discover why they are more likely to roll some numbers than others.

You can have your group use either of these handouts—or you can use both.

Experimenting with *Pig* Probabilities

To experiment with probability, your group will divide into pairs. Each pair will roll their dice at least 20 times and record their results. Working with the group and using the form on page 22, you will create a bar graph. This graph will give you a visual representation of the results, so that everyone can see which numbers are most likely to come up.

To make a meaningful graph, you need to collect the results of 200–250 rolls. So, if you have 10 pairs of experimenters and each pair rolls the dice 20 times, you'll have a good experimental sample. If you have fewer

experimenters, each pair will need to roll the dice more times.

After everyone has finished rolling, ask each pair of experimenters how many times they rolled the number 2. Add all the results, and graph the number on the bar graph on page 22. Repeat this process for each number from 3 to 12. When you are done, the graph will look something like the ones shown here.

Ask members of your group what the graph tells them. Here are two things it may reveal:

- The numbers 2 and 12 don't come up very often.

- The numbers 6, 7, and 8 come up frequently.

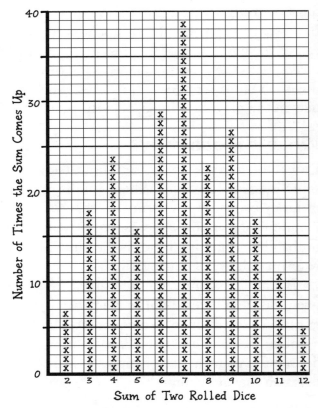

Expected Outcomes from Multiple Rolls of Two Dice This graph shows the results predicted by probability.

Experimental Results from Multiple Rolls of Two Dice This graph shows the experimental results from 216 rolls of the dice.

You can see from the graphs on page 20 that experimental results may not exactly match the expected outcomes predicted by probability theory. Instead, experimental data show *tendencies*. (Remember, probability doesn't say what *is going to* happen. It merely says what *is likely to* happen.)

Exploring *Pig* Probabilities

Looking at the *Pig Possibilities* chart can help people see that they are more likely to roll some numbers than others. Using different-colored pens or different shapes to circle all the ways to roll each number creates a visual pattern on the chart, which makes understanding the chart easier.

If you have a small group, have the members look at the chart together and work through the questions. Or have people work in pairs and figure out the answers to the questions.

Ask your group to count the possible outcomes in the chart. When a player rolls two dice, there are 36 different combinations that can come up.

You may have to point out that rolling 1 on Die A and 6 on Die B is *different* from rolling 6 on Die B and 1 on Die A. These are two different events even though they result in the same numbers. If people have trouble with this, have them imagine that the dice are different colors—a red die and a blue die, for instance. Rolling 1 on the red die and 6 on the blue die is clearly different from rolling 6 on the red die and 1 on the blue die.

Of the 36 possible outcomes, there's only one way to roll 2—by rolling 1 on Die A and 1 on Die B. But there are six ways to roll 7:

- Roll 6 on Die A and 1 on Die B.
- Roll 5 on Die A and 2 on Die B.
- Roll 4 on Die A and 3 on Die B.
- Roll 3 on Die A and 4 on Die B.
- Roll 2 on Die A and 5 on Die B.
- Roll 1 on Die A and 6 on Die B.

There are six chances that a player will roll 7, and only one chance that a player will roll 2. Rolling 7 is six times as likely as rolling 2.

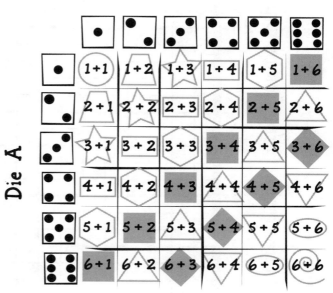

What's This Got to Do with *Pig*?

Knowing which numbers are most likely to come up is an advantage in playing *Pig*, particularly when players are very close to their target score of 100.

For example, suppose a player has 90 points. Should he or she roll again or stay put? Just 10 points are needed to reach 100. On another roll, he or she might roll 11 or 12 and lose.

To figure out whether to roll again, the player might look at the *Pig Possibilities* chart and count how many ways there are to roll 11 or 12. There are only three ways to roll 11 or

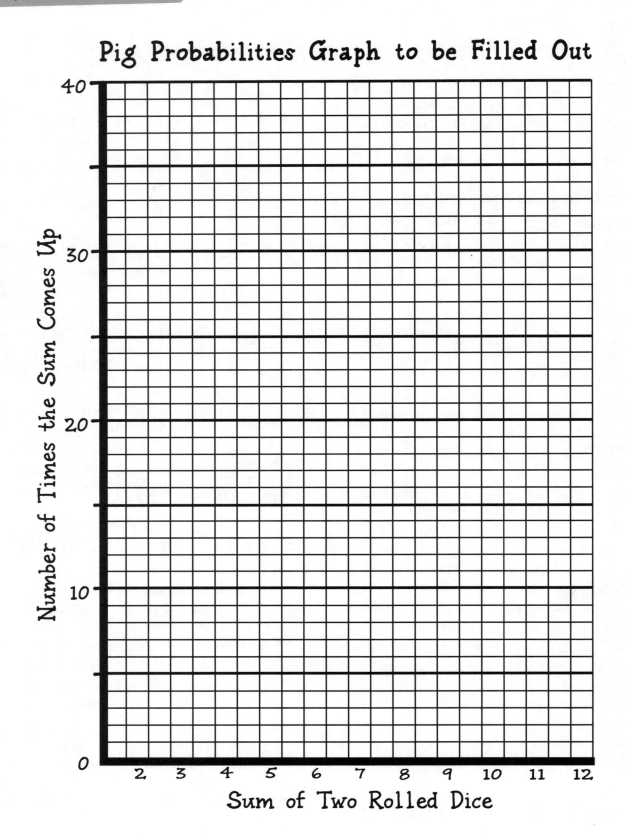

Pig Probabilities Graph to be Filled Out

Number of Times the Sum Comes Up

Sum of Two Rolled Dice

The Math Explorer
Published by Key Curriculum Press / © 2003 Exploratorium

12—meaning there are 33 ways to roll lower numbers. So the player has a very good chance of rolling again and not going over 100.

Now suppose a player has 94 points. He needs 6 points to reach 100. If he rolls any number greater than 6, he'll lose. The *Pig Possibilities* chart shows that there are 21 ways to roll a number greater than 6—and only 15 ways to roll 6 or less. Rolling with a score of 94 points looks a lot less inviting than rolling with a score of 90 points.

If your group understands this concept, you may want to show them how to figure out the probability of rolling each number.

Calculating Probability

When you flip a coin, you have an equal probability of getting heads or tails. Suppose you win if you get heads. How likely are you to win?

One way to express the probability that you will win is to say that you have one chance in two of winning. There are two possible outcomes: heads or tails. One of those outcomes means you win. So the probability that you'll win is 1 out of 2—which you can write as a fraction: $\frac{1}{2}$.

The probability that something will happen is a number from zero to 1 that measures the likelihood that the event will occur. A probability of zero means that an event will never happen. A probability of 1 means that an event is sure to happen. (There's a probability of 1 that the sun will rise tomorrow morning, for example.) A

Where's the Math?

Some middle schoolers may have been introduced to ratios in math class. A *ratio* is a comparison of two quantities. Calculating probability involves creating a ratio by comparing the number of ways a certain event can occur to the total number of equally likely possible outcomes. The probability of getting heads in a coin toss is $\frac{1}{2}$, as there is only one way to get heads out of two possible outcomes.

People sometimes say that the probability of getting heads is 50-50. How does that relate to a probability of $\frac{1}{2}$?

A fraction can be converted into a decimal number. Divide 2 into 1, and you get 0.5, which is the same as $\frac{1}{2}$. To convert this to a percent, multiply 0.5 by 100. You get 50 percent. That is, heads will turn up 50 percent of the time, which is why people say the chances are 50-50.

probability of $\frac{1}{2}$ means that an event is likely to happen half the time.

There are more possible outcomes when you roll a pair of dice. You have to take that into account when you calculate the probability that a certain number, or a certain set of numbers, will come up when you roll the dice.

To calculate the probability that a player will roll 10 or less, follow these steps:

Step 1 Count how many outcomes are possible when the dice are rolled. (Counting all the boxes in the *Pig Possibilities* chart reveals that there are 36 possible equally likely outcomes.)

Step 2 Count how many outcomes result in 10 or less. (There are 33 such outcomes.)

Step 3 Create a fraction by putting the number of outcomes that give you numbers greater than 10 over the total number of possible outcomes. (Put the number from Step 2 over the number from Step 1.) The fraction is $\frac{33}{36}$. That's the probability of rolling 10 or less. If your group understands fractions, it can figure out that $\frac{11}{12}$ is the same as $\frac{33}{36}$.

What does a probability of $\frac{11}{12}$ mean? It means that if a player rolls the dice 12 times, he or she is likely to roll 10 or less 11 times. That means the chances of rolling 10 or less are pretty good. Remember: a probability of zero means that an event will never happen, and a probability of 1 means that an event is certain to happen. The fraction $\frac{11}{12}$ is much closer to 1 than to zero. So it's likely—though not certain—that you will roll 10 or less.

What's Next?

Can group members figure out the probability of rolling 6 or less? Can they figure out the probability of rolling 2, or "snake eyes"? To calculate either probability, follow the procedure above.

The Math Explorer
Published by Key Curriculum Press / © 2003 Exploratorium

Madagascar Solitaire

Madagascar Solitaire is an ancient game from the island country of Madagascar, just off the southeast coast of Africa. This game encourages players to think about strategy and symmetry. It also teaches skills related to graphing.

Madagascar is also known as the only place in the world where lemurs live in the wild.

Preparation and Materials

For each player, you will need:

- a handful of stones, beans, pennies, paper clips, or other small objects to use as markers

- copies of *Simple Madagascar Solitaire, Simple Madagascar Solitaire Score Card, Intermediate Madagascar Solitaire,* and *Intermediate Madagascar Solitaire Score Card*

- a copy of *Advanced Madagascar Solitaire* (optional)

You will also need:

- copies of *Playing Madagascar Solitaire* (1 per group of 3 or 4)

- an extra copy of *Intermediate Madagascar Solitaire*

Using This Activity

Tips for using *Madagascar Solitaire* with a group start on page 32. You can also have people play individually. Simply give them a handful of markers and copies of the instructions, game boards, and score cards. Tell them to follow the instructions.

If your group has "homework hour," playing this game is a good activity for those who finish their work early.

Planning chart

Simple game	about 10 minutes
Intermediate game	about 35 minutes
Advanced game	open-ended

Playing Madagascar Solitaire

Here's a game of solitaire from Madagascar, an island country off the southeast coast of Africa. Games like this are found all over the world.

The object of the game is to remove all the markers from the game board one by one by jumping over them with other markers. You win when you have only one marker left.

What Do I Need?

◇ a handful of stones, beans, paper clips, or other small objects to use as markers

◇ copies of *Simple Madagascar Solitaire* and *Intermediate Madagascar Solitaire*, and score cards for both games

The island of Madagascar is home to the ring-tailed lemur.

Here's How to Play

1 Put a marker in every circle of the game board.

2 Remove one marker from your board.

3 Jump one marker over another. You can jump over only one marker at a time. You have to jump over a marker. (No fair jumping an empty space.) You must land in an empty circle.

4 After you jump, remove the marker you jumped over. Then jump again—if you can. You can make your next jump with any marker, and not only the one you just used.

5 If you end up with only one marker on the board, you win! If you have more than one marker left, and no jumps are possible, you lose!

The Math Explorer
Published by Key Curriculum Press / © 2003 Exploratorium

Simple Madagascar Solitaire

Use this easy game to get ready for *Intermediate Madagascar Solitaire*.

Game Board

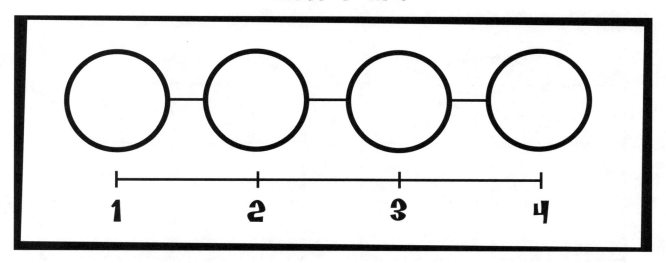

Step 1 Play a game. Keep track of your moves on the *Simple Madagascar Solitaire Score Card*. Did you win? If you won, put a star on your score card beside that game. If you lost, don't put a star.

Step 2 Try playing again, starting with a different marker. Can you figure out which marker you must remove first in order to win?

Step 3 After you find one winning first move, try to find another. Always put a star by a game that you win. How many different winning first moves can you find?

Step 4 Are the winning moves similar to each other in any way?

Simple Madagascar Solitaire Score Card

When you are playing *Simple Madagascar Solitaire*, write down the position of each marker you move on this score card. It will help you figure out which moves lead to a winning game.

Game 1

Marker Moved	New Position
	removed

Game 2

Marker Moved	New Position
	removed

Game 3

Marker Moved	New Position
	removed

Game 4

Marker Moved	New Position
	removed

The Math Explorer
Published by Key Curriculum Press / © 2003 Exploratorium

Intermediate Madagascar Solitaire

This game has more possible moves than *Simple Madagascar Solitaire*. Markers can jump up and down or left and right, but they can't jump on the diagonal.

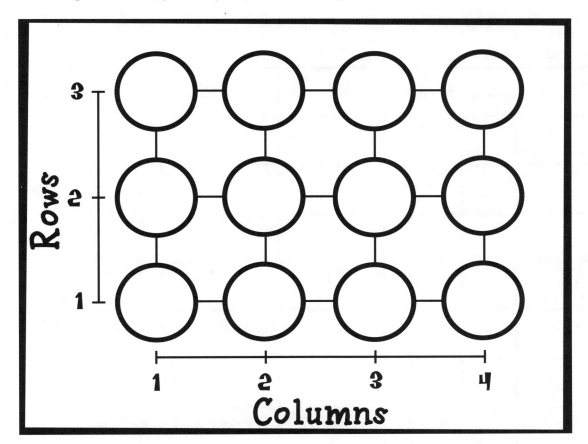

Step 1 Play a few games. Keep track of your moves on the *Intermediate Madagascar Solitaire Score Card*. When you win a game, put a star by that game on your score card.

Step 2 This game board is *symmetrical*. That means you can fold it so that the circles on one half match up with the

circles on the other half. Try it. The fold marks a *line of symmetry*. The board on one side of this line is a *mirror reflection* of the board on the other side. Find two lines of symmetry for this board.

Step 3 After you win a game, try to win by starting with a mirror reflection of the first move in your winning game.

Explorer's Name: _____   Date: _____

Intermediate Madagascar Solitaire Score Card

To keep track of your moves, write the column and the row of the marker you move, and the column and the row of its new position.

Marker Moved		New Position		Marker Moved		New Position	
Column	Row	Column	Row	Column	Row	Column	Row

When you write down a column and row, you are using two numbers to identify an exact position. In math class, teachers talk about (x, y) coordinates. On the game board, the column number is the x-coordinate and the row number is the y-coordinate.

The Math Explorer
Published by Key Curriculum Press / © 2003 Exploratorium

Advanced Madagascar Solitaire

You can play *Madagascar Solitaire* on many different game boards.

Step 1 Draw your own game board. You might want to try a board with four columns and four rows or a board with six columns and six rows.

Step 2 Play a game on whatever new board you've chosen. *Keep an open mind, and have fun!*

Step 3 What lines of symmetry can you find on your new game board?

Step 4 For an extra challenge, use the game board below, which is used in Madagascar.

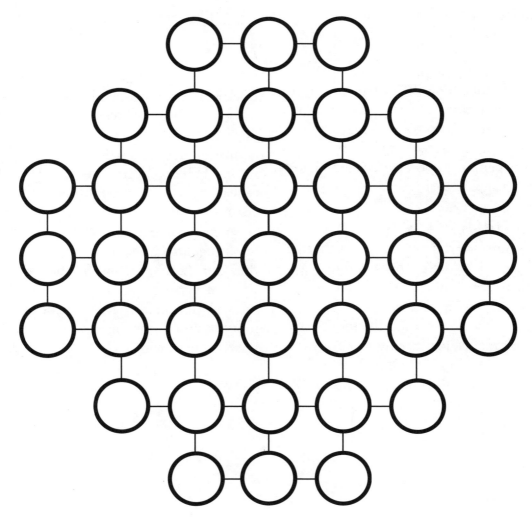

Using *Madagascar Solitaire* with a Group

This is a great game to use with a group in which skill levels vary. By using different game boards, you can make playing this game easier or more difficult. Because this is a solitaire game, players can begin with a simple version and move on to more challenging ones when they are ready.

Playing *Madagascar Solitaire*

Ask your group if anyone knows any solitaire games. Students new to English may not know the word *solitaire*, though they may be familiar with such games.

You can have people play this game individually by following the instructions (on page 26). Or you can use *Madagascar Solitaire* as a group activity. To do so, we suggest you divide your group into smaller clusters of three or four people. Give each cluster a copy of *Playing Madagascar Solitaire*.

Keeping Track of Moves

The instructions for the game suggest that people keep track of their moves as they play. This lets them show others how they won. When they are playing *Intermediate Madagascar Solitaire*, recording the moves also gives people practice identifying locations, using an (x, y) coordinate system. This will come in handy when they make graphs in math class. (See page 38.)

Some people find it frustrating to keep track of their moves while they are playing the game. Eva Jo Meyers at the Boys and Girls Clubs of San Francisco simply let group members play for a while, and they really got into it. When players won, she asked them to repeat the moves to see if they could do it again.

After people have won, they have much more incentive to record their moves—so that they can show everyone else how they did it! Even players who aren't recording all their moves will find it useful to record their first move. Certain starting moves make it impossible to win!

Simple Madagascar Solitaire

Have everyone play this very simple game. When someone wins, ask this person to show you how he or she won.

Whether a player wins this game depends on which marker the player takes away first. Look at the game board below. A player will always lose if he or she starts the game by removing the marker in Position 1 or Position 4. A player will win if he or she first removes the marker in Position 2 or Position 3.

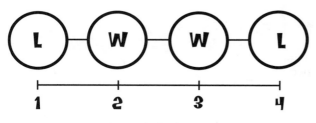

L stands for Loser.
W stands for Winner.

Introducing the Concept of Symmetry

The last step of *Simple Madagascar Solitaire* on page 27 asks the question "Are the winning moves similar to each other in any way?" This question gives you an opportunity to introduce the concept of *symmetry* to your group.

The two winning moves are actually the same if you look at this game mathematically. The two halves of the board are *symmetrical*. That means that when you fold the *Simple*

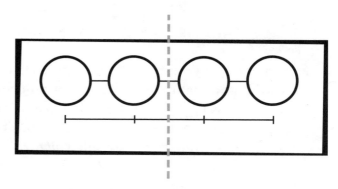

The dashed line marks the line of symmetry

Madagascar Solitaire game board between Positions 2 and 3, everything on the left side of the fold has an exact match on the right side of the fold.

This fold (shown by a dashed line in the illustration) marks a *line of symmetry*. This kind of symmetry is *mirror symmetry* or *reflection symmetry*: the left half of the board is a mirror reflection of the right half.

In terms of winning the game, removing the marker from Position 2 has the same effect as removing the marker from Position 3. The moves are mirror images of each other. And the same is true for Positions 1 and 4.

Simple Madagascar Solitaire is such an easy game that looking for the line of symmetry doesn't help you much. But finding lines of symmetry can be a big help when the group moves on to *Intermediate Madagascar Solitaire*.

Where's the Math?

Symmetry is an important concept in science, art, and mathematics. Many things in nature have mirror, or reflection, symmetry. Your face, for instance, is probably more or less symmetrical, with a line of symmetry running down the middle of your face between your eyes. This type of symmetry is also called *bilateral symmetry* (*bi* comes from the Latin word for "two," and *lateral* comes from the Latin word for "side"). In middle school mathematics, students may have to look for lines of symmetry in various geometric figures.

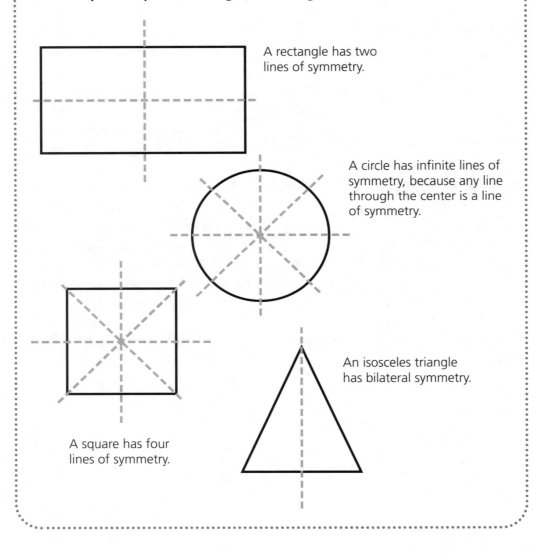

A rectangle has two lines of symmetry.

A circle has infinite lines of symmetry, because any line through the center is a line of symmetry.

A square has four lines of symmetry.

An isosceles triangle has bilateral symmetry.

Intermediate Madagascar Solitaire

Before people start playing *Intermediate Madagascar Solitaire*, you may want to talk about strategy. Strategy involves looking ahead and thinking about the consequences of each move. Suggest that players ask themselves, "If I jump here, how will that affect what I can do next?" Thinking ahead will help them win more often.

The point is not merely to win the game, but also to be able to tell other people what starting move led to the winning game. Challenge the players to win a game—and then to show how they won.

Some people may want to work in pairs, with one person playing the game and the other keeping track of the moves. Players can use the *Intermediate Madagascar Score Card* to track their moves. The chart to the right gives an example of how to do this.

Other players may prefer to play until they win—and then repeat the winning game while recording their moves.

For people who have a hard time getting started, Sharah Nieto of the Boys and Girls Clubs of San Francisco suggests drawing a game board on the chalkboard and using sticky notes for markers, so the group can play a game together.

Looking for Winning Moves

In *Simple Madagascar Solitaire*, the wrong first move can ensure that a player will lose the game. The same is true in *Intermediate Madagascar Solitaire*. Because the game is more complex, it's harder to find all the moves that can lead to a winning game.

Can your group, working together, identify all the starting moves that make it

Marker Moved		New Position	
Column	Row	Column	Row
4	1	removed	
2	1	4	1
2	3	2	1
4	3	2	3
4	2	2	2
1	1	3	1
4	1	2	1
1	3	1	1
1	1	3	1
2	3	2	1
3	1	1	1

Win!
Here's one winning game.
Can your group find others?

possible to win? This may take a while, but it's fun to try.

Ask players to tell you when they win—and to tell you the first move they made in their winning game. Using the extra copy of the *Intermediate Madagascar Solitaire* game board, keep track of which starting moves lead to winning games. When someone wins a game, mark a *W* in the circle that was his or her first move.

Be aware that it's possible to lose a game even if you start with a winning first move. The first move of a losing game is not always a losing move.

Finding Symmetry

After people have won a few games, you may want to point out that symmetry can help them find other ways to win.

If people discussed symmetry when they played *Simple Madagascar Solitaire*, remind them of that discussion. The *Simple Madagascar Solitaire* game board has one line of symmetry. Folded on this line, all the circles on one side of the fold match up with all the circles on the other.

Challenge players to figure out how they can fold the *Intermediate Madagascar Solitaire* game board to find the lines of symmetry. There are two lines of symmetry.

Winning with Symmetrical Moves

How can finding lines of symmetry help win a game?

Take a look at the *Simple Madagascar Solitaire* game board pictured on page 27. If you imagine that the line of symmetry is a mirror, then the two winning first moves are mirror images of each other.

The same is true of winning moves in *Intermediate Madagascar Solitaire*. A player can win if the first move is to remove the marker in Position 4, 1. (See the chart on page 35.) Imagine that one of the lines of symmetry on the *Intermediate Madagascar Solitaire* game board is a mirror, and imagine the mirror image of that winning first move. (See the illustration at the bottom of this page.)

A player can win by starting with this move and playing a game in which every move is a mirror image of the move in the winning game! Challenge your group to play a winning game starting with moves that are mirror images of winning first moves.

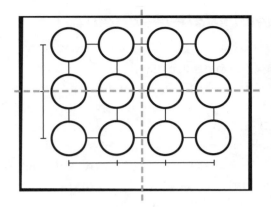

The dashed lines mark the lines of symmetry.

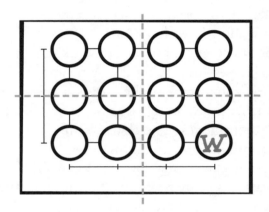

This is a winning first move . . .

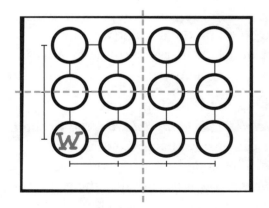

. . . and this is one mirror image of that move. Can you find another mirror image of this move?

Advanced Madagascar Solitaire

If your group is having a good time playing *Madagascar Solitaire*, have players design their own game boards to play more advanced versions of the game.

Other Game Boards to Try

Here are some game boards for players who want to try a more challenging game:

- a board with four columns and four rows
- a board with six columns and six rows
- the version played in Madagascar, shown on page 31

Looking for Symmetry

Players may find many lines of symmetry, depending on the game board. They can always test for symmetry by folding the paper. If all the circles on one side of the fold match up with circles on the other side, then that fold is a line of symmetry.

Rotational Symmetry

If you like, you can introduce your group to a different type of symmetry. Suppose you take a copy of the game board on page 31 and rotate it one quarter turn. The rotated game board will fit perfectly on top of the original. Rotate it another quarter turn, and it will fit again. This is what mathematicians call *rotational symmetry*.

A game board has rotational symmetry if there is a point on the image around which the board can be turned a certain number of degrees (but less than 360 degrees) and look the same as the original.

A good example of something familiar that has rotational symmetry is a pizza that's been cut into slices. Suppose all the slices are the same size and have the same ingredients in the same places. If you turn the pizza just the right amount, it will look the same as it did before you turned it.

If a player finds a winning move on a board that has rotational symmetry, he or she can win again by finding the rotationally symmetric, equivalent move.

Where's the Math?

This activity encourages people to explore the concept of symmetry. It also offers practice in two important mathematical skills: solving problems by reasoning and recording data using ordered pairs.

Solving Problems To win these solitaire games, players must think strategically, looking ahead, figuring out the consequences of each move, and searching for patterns that will lead to a winning game. This sort of reasoning is at the heart of mathematical thinking.

Using Ordered Pairs When students begin graphing in math class, they learn about *ordered pairs*. When players write down a column and row of the game board in *Intermediate Madagascar Solitaire*, they are writing down an ordered pair. Saying the pair is "ordered" means that the order in which the numbers are written is important. In this case, the number of the column precedes the number of the row. If the order of the numbers is switched, the numbers will identify a different location on the board.

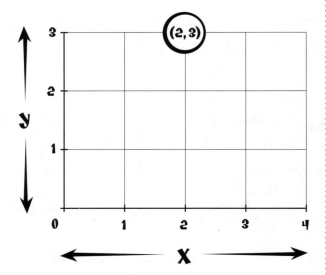

The (x, y) coordinate (2, 3) represents a particular location on this graph. If the numbers were reversed, they would identify a different location.

In graphing, people also talk about (x, y) *coordinates*. The pair of coordinates (x, y) is an ordered pair; x is always written first. In *Madagascar Solitaire*, the column is the x-coordinate and the row is the y-coordinate.

Fantastic Four

Playing *Fantastic Four* is sure to sharpen players' skills in basic mathematics and in forming equations. This game also helps develop problem solving skills. Players often improve rapidly, seeing more strategies each time they play.

Preparation and Materials

For each group of two to five players, you will need:

- a deck of cards
- a timer or clock
- pencils and scratch paper
- a copy of *Playing Fantastic Four*

Using This Activity

Tips for how to lead a group in playing *Fantastic Four* start on page 42. *Fantastic Four* can also be played independently by groups of two to five people—simply give them a copy of the instructions and a deck of cards. Groups can play this game over and over again. Each time they play, the numbers will be different, presenting a new challenge.

If you have an hour or more, you can combine this game with one or more other games.

Planning chart

Introducing *Fantastic Four*	10 minutes
Playing *Fantastic Four*	20 minutes or longer

Playing Fantastic Four

Can you multiply, divide, add, or subtract the numbers on
the bottom four cards to make the number on the
fifth card? In *Fantastic Four,* you win the game by
forming the most equations using the most numbers.

What Do I Need?

◊ a deck of cards

◊ a timer or clock

◊ a pencil or pen

◊ a few people to play with

◊ paper to write on

Here's How to Play

1 Remove the face cards (jacks, queens, and kings) from the deck,
and set them aside.

2 One player deals out four cards face up and one card face down.

3 On a piece of paper, each player writes the four numbers that
are face up.

4 If you have a timer, set it for five minutes. If you are using a clock,
have someone else let the players know when five minutes is up.

5 When everyone's ready, start the timer and flip over the fifth card.

6 For five minutes, each player creates and writes down equations
that use some or all of the numbers on the first four cards that
yield the number on the fifth card.

7 The goal is to write as many equations as possible. You earn the
most points for equations that use all four numbers to make the
fifth number (and no points for using only one number).

Here's an Example!

Suppose the four cards you turn over are 4, 9, ace (which is 1), and 2. The fifth card is a 7. Can you make 7 using 4, 9, 1, and 2?

Here are three ways to do it:

$$9 - \frac{4}{2} = 7 \qquad\qquad 4 + 2 + 1 = 7$$

$$\frac{9 - (4 - 2)}{1} = 7$$

Can you think of any other ways?

How can I make 7 using 4, 9, 1, and 2? 9 - 2 = 7! That's one way.

Here's How to Score

1 When the five minutes are up, all players stop working and figure out their scores.
   - Equations using two of the four cards score 4 points.
   - Equations using three of the four cards score 9 points.
   - Equations using all four cards score 16 points!

2 Each player checks all the other players' equations. It's OK if some of you have the same equations. If a player made a mistake, the person who finds the mistake earns 7 points—and the player who made the mistake gets no points for that equation.

The highest score wins!

Playing *Fantastic Four*

This game gives people practice in mathematical operations, such as addition, subtraction, multiplication, and division. It also encourages them to write equations correctly, including making use of parentheses. It can be very competitive—and a lot of fun.

Introducing the Game

We suggest that you introduce this game by showing the group how to play it. (See "Here's How to Play" on page 40.) Deal out four cards, and then a fifth, and ask the group if it can form an equation that uses some or all of the numbers on the four cards to equal the number on the fifth. (See page 41 for an example.)

Give people a few minutes to jot down possibilities on their own. Then ask what their equations are. Write the equations for everyone to see.

Try to show examples of equations using two numbers (worth 4 points), three numbers (worth 9 points), and all four numbers (worth a whopping 16 points). Don't worry if you can't find an equation that uses all four numbers. Sometimes it just isn't possible.

If people have trouble coming up with equations, introduce the helpful strategies described at right, and on page 43.

Playing the Game

Once players understand the rules, they can play this game in groups of two to five. Or you can have the entire group play together. In a large group, the leader deals out the cards and keeps track of the time. Once time is up, players calculate their own scores. Ask who has the highest score, and then have the group check that person's equations. Or have players swap papers and check each other's equations. Remind everyone that a player earns 7 points for finding someone else's mistake!

A Word About Operations and Parentheses

Addition, subtraction, multiplication, and division are *mathematical operations*. So are square roots, exponentials, and factorials. (See the glossary for definitions of these terms). After playing *Fantastic Four* using addition, subtraction, multiplication, and division, your group may decide to include other operations.

A series of operations can yield different answers, depending on the order in which the operations are done. That's why it's important for players to use parentheses when they write their equations. In a series of operations, the operations inside parentheses are executed first. People should use parentheses to make it clear which operations to do first. (For more about this, see "Where's the Math?" on page 43.)

Helpful Strategies

If people are having trouble coming up with equations that work, it may help to focus on finding combinations that lead to ones or zeros.

Eliminating Numbers by Making Zero

Suppose you get stuck trying to use 3, 4, 7, and 7 in an equation to obtain 4. You can "get rid" of some numbers by creating a zero, like this:

$$\frac{7-7}{3} + 4 = \frac{0}{3} + 4 = 0 + 4 = 4$$

Notice how the first term disappears. Dividing anything *into* zero gives zero. You

The Math Explorer
Published by Key Curriculum Press / © 2003 Exploratorium

may have to remind people that they can divide *into* zero but not *by* zero. So, they can't have

$$\frac{3}{7 - 7}$$

because zero can never be on the bottom of a fraction (in the denominator).

Eliminating Numbers by Making 1

Suppose you're trying to use 3, 4, 7, and 7 in an equation to obtain 7. You can get rid of unwanted numbers by turning them into 1, like this:

$$\frac{3 + 4}{7} \times 7 = \frac{7}{7} \times 7 = 1 \times 7 = 7$$

Where's the Math?

When performing a series of calculations, it's important to do them in the right order. Here are two mathematical expressions that look very similar:

$$(1 + 2) \times 3$$
$$1 + (2 \times 3)$$

The numbers and the operations are the same—but the answer depends on which operation is done first.

In doing any calculation, operations within parentheses are done first. The answer to the first calculation is found this way:

$$(1 + 2) \times 3 = 3 \times 3 = 9$$

And the answer to the second calculation is found this way:

$$1 + (2 \times 3) = 1 + 6 = 7$$

Very different answers!

Be sure people in your group understand this and use parentheses to identify the order in which they want operations done.

Mathematicians have a set of rules governing the order in which operations are done:

- First, do whatever is inside parentheses.
- If there are any exponents, do those next.
- Then do all the multiplications and divisions, working through the operations in the order they appear from left to right.
- Finally, do all additions and subtractions from left to right.

To remember the order of operations, people can memorize the sentence *Please Excuse My Dear Aunt Sally.* The first letters of the words reveal what to do:

Please	**P**arentheses
Excuse	**E**xponents
My Dear	**M**ultiply and **D**ivide
Aunt Sally	**A**dd and **S**ubtract

Fast Times

Many middle schoolers still have trouble with the multiplication table. As a result, they struggle with problems requiring this basic knowledge. Knowing the multiplication table is essential to working effectively with fractions and to preparing for algebra.

In this section, you'll find one activity and two games. *Eratosthenes' Sieve* gives people a new view of the multiplication table, letting them discover hidden patterns in the numbers from 1 to 100. This review of multiplication will come in handy when your group plays *Hopping Hundred* and *Tic-Tac-Toe Times*, games that require a knowledge of the multiplication table.

The material in this section offers a fun way to help members of your group improve their basic math skills.

Activity 6 ## Eratosthenes' Sieve

Using a technique developed by an ancient Greek athlete and mathematician, people will discover patterns that help them understand the multiplication table in a new way.

Activity 7 ## Hopping Hundred

This fun, easy-to-play game gives players a chance to practice skills that are essential in working with fractions.

Activity 8 ## Tic-Tac-Toe Times

While engaged in this game, players will also be reviewing the multiplication table.

Eratosthenes' Sieve

Many middle schoolers struggle with the multiplication table and problems that require this basic knowledge. This activity will help those who have difficulty with multiplication.

This activity also introduces prime numbers. A prime number is a number that can be divided evenly only by 1 and itself. "Divided evenly" means that the result is a whole number, with no remainder. Prime numbers and their properties were first studied extensively by ancient Greek mathematicians, including Eratosthenes (pronounced AIR-a-TOSS-the-knees).

Preparation and Materials

Each member of your group will need:

- a pencil
- 4 different-colored pens or markers (yellow, red, green, and blue)
- a copy of *Hundreds Chart*

This activity works best if you lead your group through the steps, using *Eratosthenes' Sieve* as your guide. If you want people to work independently, each will need a copy of *Eratosthenes' Sieve*.

Using This Activity

By completing *Eratosthenes' Sieve*, people will acquire a tool to help them play *Hopping Hundred* (page 55) and *Tic-Tac-Toe Times* (page 61), games that require knowledge of the multiplication table. You may want to do this activity right before playing one or both of those games.

Tips for how to use *Eratosthenes' Sieve* start on page 53.

Planning chart

Finding the prime numbers by using *Eratosthenes' Sieve*	20 minutes

The Math Explorer
Published by Key Curriculum Press / © 2003 Exploratorium

Eratosthenes' Sieve

If you've ever made cookies, you may have used a sieve to sift flour. A *sieve* is a device with a bottom that's full of tiny holes. It lets fine flour particles through but retains the lumpy bits.

Eratosthenes (pronounced AIR-a-TOSS-the-knees) was an ancient Greek athlete, astronomer, and mathematician. His sieve doesn't sort out lumps of flour. It sorts out prime numbers. A *prime number* is a number that can be divided evenly only by 1 and itself. "Divided evenly" means that the result is a whole number, with no remainder.

What Do I Need?

◇ a pencil

◇ a *Hundreds Chart* (page 52)

◇ 4 different-colored pens or markers (yellow, red, green, and blue)

What Do I Do?

When you mark your *Hundreds Chart* according to the steps that follow, you'll reveal all the prime numbers in the chart. At the same time, you'll discover patterns that will help you remember the multiplication table.

Step 1 Using your pencil, draw an *X* through the number 1 on your *Hundreds Chart*. The number 1 is a special number. It's not prime, because a prime number has to be evenly divisible by two numbers: 1 *and* itself. The number

1 is divisible by only one number: 1. So, it's not prime. (That may seem picky, but mathematicians are like that sometimes.)

Step 2 Look at the number 2. The number 2 is prime because only two numbers can divide into it evenly: 1 and 2. Circle the 2 with your pencil.

Step 3 Using your yellow marker, draw a vertical line through the center of the box the number 2 is in.

Step 4 Find all the other multiples of 2 in the chart, and draw the same yellow vertical line through each of them. Is 3 a multiple of 2? How about 4? The number 4 is a multiple of 2 because it equals 2 × 2.

To find the multiples of 2, look for the *even* numbers. Be sure to draw a yellow vertical line through every even number between 2 and 100. Do you see a pattern in the multiples of 2?

Step 5 At the bottom of your chart, you'll find a place to create a key to help you remember what the different colors indicate. Draw a yellow vertical line through the box labeled "Multiples of 2." Any number that has a yellow vertical line through it is a multiple of 2.

Step 6 Look at the next number in the top row: 3. Is 3 a prime number? It is, because it can be divided evenly only by 1 and 3. Circle the 3 with your pencil.

Step 7 Using your red marker, draw a diagonal line across the box containing the number 3. Make the diagonal from the bottom left corner to the top right corner.

The Math Explorer
Published by Key Curriculum Press / © 2003 Exploratorium

Step 8 Using the red marker, draw the same diagonal through all the multiples of 3 in the chart. The number 6 is a multiple of 3; it is 3 multiplied by 2. Find all the multiples of 3 between 3 and 100. After you've marked several multiples of 3, you may be able to find a pattern to help you locate the others.

I know 81 is a multiple of 3 because 8 + 1 = 9. Because 9 is a multiple of 3, so is 81!

There is also a trick for finding the multiples of 3. Add up the digits, and if that sum is a multiple of 3, then the number itself is a multiple of 3.

Step 9 In the key at the bottom of your chart, draw a red diagonal line through the box labeled "Multiples of 3."

Notice that some boxes in your chart have both a yellow vertical line and a red diagonal line. These numbers are multiples of both 2 and 3. If a number is a multiple of two numbers, it's also a multiple of those two numbers multiplied together! So, these boxes show you the multiples of 2 × 3, or 6. Mark this on your key, too, by drawing both a yellow vertical line and a red diagonal line in the box labeled "Multiples of 6."

Step 10 Continuing along the top row of the chart, notice that 4 is already marked with a yellow vertical line, reminding you that it's a multiple of 2, so it can't be a prime number. The next number, 5, is prime, so circle it with your pencil.

Using your green marker, draw a vertical line through the 5. Draw this green line a little left of center so that you don't cover up any of the yellow lines you already drew in the multiples of 2.

Step 11 Find all the multiples of 5 in the chart, and draw the same green line through each of them. The trick to finding multiples of 5 is to look for numbers that end in 5 or zero. Look for a pattern that will help you find all the multiples of 5. In your key, draw a green line through the box labeled "Multiples of 5."

Step 12 Looking at your key and your chart, can you figure out what set of colored lines shows you the multiples of 10? (Here's a hint: $2 \times 5 = 10$.) Now look for multiples of 15 (which is 3×5) and 30 (which is $3 \times 5 \times 2$). Add these to your key.

Step 13 The number 6 has already been marked. The number 7 is the next prime. Circle 7 with your pencil. Using your blue marker, draw a diagonal line across 7. This time, draw the diagonal from the top left corner to the bottom right corner so that it doesn't cover up the diagonal you drew for multiples of 3.

Step 14 Find all the multiples of 7 in the chart, and draw a blue diagonal line through each of them. There is no trick for finding multiples of 7, but these numbers do form a pattern on your chart. Look for the pattern.

The Math Explorer
Published by Key Curriculum Press / © 2003 Exploratorium

Step 15 In your key, draw a diagonal blue line through the box labeled "Multiples of 7." If you want to, you can add other boxes to your key. Can you figure out what set of colored lines shows you multiples of 14? Or 21? Or 35?

Step 16 Continuing along the chart, notice that 8, 9, and 10 are already marked. There are no prime numbers left in the top row. All the numbers that haven't been marked yet are the remaining primes between 1 and 100. Circle all the remaining numbers.

Step 17 How many prime numbers are there between 1 and 100? You should get the same answer as everyone else in your group.

Hundreds Chart

1	2	3	4	5	6	7	8	9	10
11	12	13	14	15	16	17	18	19	20
21	22	23	24	25	26	27	28	29	30
31	32	33	34	35	36	37	38	39	40
41	42	43	44	45	46	47	48	49	50
51	52	53	54	55	56	57	58	59	60
61	62	63	64	65	66	67	68	69	70
71	72	73	74	75	76	77	78	79	80
81	82	83	84	85	86	87	88	89	90
91	92	93	94	95	96	97	98	99	100

☐ Multiples of 2   ☐ Multiples of 5   ☐ Multiples of 30

☐ Multiples of 3   ☐ Multiples of 10   ☐ Multiples of 7

☐ Multiples of 6   ☐ Multiples of 15

Published by Key Curriculum Press / © 2003 Exploratorium

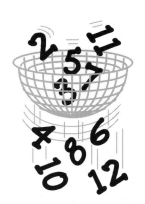

Leading Your Group Through *Eratosthenes' Sieve*

By taking a different approach, this activity may help people who have never managed to memorize the multiplication table. Mastery of the multiplication table is important for middle schoolers because this knowledge is essential to working effectively with fractions and doing algebra.

About Eratosthenes

Eratosthenes, the Greek mathematician who developed this method of finding prime numbers, was born in 276 B.C. in Cyrene, which is now in Libya in North Africa. He is the first person known to have calculated Earth's circumference.

Getting Ready

This activity is easiest to lead if everyone has pens or markers in the same four colors. The instructions assume that everyone has yellow, red, green, and blue, but you can adjust if you need to.

Before your group starts, make sure the members know the difference between odd numbers and even numbers. You should also discuss multiples. A multiple is what results when you multiply a number by other numbers. Some multiples of the number 2, for instance, are 4 (which is 2 × 2), 6 (which is 2 × 3), and 14 (which is 2 × 7).

The best way to present this activity is to demonstrate each step and have people follow along, step by step. We suggest that you provide a copy of *Hundreds Chart* (page 52) to each member of your group and lead the group through the activity using the *Eratosthenes' Sieve* instructions (page 47) as your guide. If you want people to work individually, you can give each a copy of the *Eratosthenes' Sieve* instructions as well.

Finding Patterns

Here are some of the patterns that group members may notice as they work through the activity:

- All multiples of 2 are even numbers.
- All multiples of 2 are lined up in vertical columns on the chart.
- Multiples of 3 form diagonals in the chart.
- Multiples of 5 are lined up in vertical columns under the numbers 5 and 10.
- Multiples of 7 form a pattern that resembles the way the knight moves in chess. Start with a multiple of 7, and move two rows down and one column to the right. There you'll find another multiple of 7.

Tricks to Remember

While marking their charts, people can learn a few tricks that will help them with multiplication:

- The trick to finding multiples of 2 is to look for even numbers.
- The trick to finding multiples of 3 is to add up the digits. If the sum is a multiple

of 3, then the original number is a multiple of 3, too.

- The trick to finding multiples of 5 is to look for numbers ending in 5 or zero.

Marking the Key

Make sure everyone updates the key at the bottom of the chart each time a new set of multiples is marked.

The key gets more interesting when people start marking multiples of composite numbers. (Remember, a *composite number* is a number found by multiplying two or more numbers together.) Once a chart is marked with all the multiples of 2 and 3, it's easy to find all the multiples of 6. Because 6 is equal

to 2×3, any box that's marked as a multiple of 2 *and* a multiple of 3 is a multiple of 6.

If a number is a multiple of two prime numbers, it's also a multiple of those two numbers multiplied together!

Counting the Prime Numbers

After people have filled out the top row of the chart, all the remaining unmarked numbers are prime. Tell everyone to circle all the remaining numbers in pencil and count them.

People may get different answers. Sometimes they miss a multiple. For example, it's easy to overlook 91 as a multiple of 7. See if everyone can agree on the number of primes between 1 and 100. The correct number is 25.

Where's the Math?

Understanding prime and composite numbers helps people when they are working with fractions.

When students are adding or subtracting fractions, they often have to find a *common denominator*—a multiple that two numbers have in common. Suppose your group had to add $\frac{2}{5}$ and $\frac{1}{6}$. One look at the completed charts reveals that 30 is the smallest number that is a multiple of 5 and of 2 and 3 (or 6). So, they would know that 30 is a common denominator of 5 and 6.

Another common fraction operation is writing a fraction in its lowest terms. That

means finding the greatest number that will divide evenly into both the *numerator* (the top part of the fraction) and the *denominator* (the bottom part of the fraction). For example, if you asked group members to write $\frac{12}{30}$ in lowest terms, they would need to find the greatest number that divides evenly into both 12 and 30. The completed chart shows that 12 and 30 are both multiples of 6. Dividing 6 into both the numerator and the denominator turns $\frac{12}{30}$ into $\frac{2}{5}$. By knowing which numbers are prime, they would know that $\frac{2}{7}$ and $\frac{12}{23}$ cannot be reduced.

Hopping Hundred

Hopping Hundred is a fun game for two people. It's simple to play, and it gives players a chance to practice multiplication and division.

I know the trick to winning *Hopping Hundred*. Do you?

Preparation and Materials

Each pair of players will need:

- a copy of *Playing Hopping Hundred*
- a copy of *Hopping Hundred Game Board* (Copy the two parts and tape the pages together.)
- 100 small objects (such as dried beans, pennies, paper clips, or sticky notes) to use as tokens

Using This Activity

Tips for how to lead your group in playing *Hopping Hundred* begin on page 59. If you have an hour or more, you can combine this game with other games.

The game can be simplified by using only the numbers 1–50 (the top half of the game board). You may want to start with this simpler version of the game, letting players move up to the version with 100 numbers when they are ready.

Hopping Hundred can also be played independently by groups of two. It is a great activity for those who finish other assignments early.

Planning chart

Playing *Hopping Hundred*   20 minutes

Playing Hopping Hundred

Hop from one number to another, but don't get stuck!

I know the trick to winning *Hopping Hundred*. Do you?

What Do I Need?

◇ a partner

◇ a *Hopping Hundred Game Board*

◇ 100 pennies, dried beans, paper clips, scraps of paper, or other small objects to use as tokens

Here's How to Play

1 Player 1 chooses any even number and puts a token on that number.

2 Player 2 chooses any number (even or odd) that is a *multiple* or a *factor* of Player 1's number and puts a token on that number. For example, suppose Player 1 chooses 10. Player 2 could choose 20, 30, or 40. These numbers are all *multiples* of 10, because you can multiply 10 by some other number to make them. Or Player 2 could choose 1, 2, or 5. These numbers are all *factors* of 10, because they divide evenly into 10. (*Divide evenly* means that the result is a whole number and there is no remainder.)

3 Players take turns choosing numbers to cover from those remaining. On each turn, a player can choose any uncovered number, even or odd, as long as it is either a multiple or a factor of the previous number chosen.

4 The first person who cannot cover a number loses the game!

The Math Explorer
Published by Key Curriculum Press / © 2003 Exploratorium

Hopping Hundred Game Board

5	10	15	20	25	30	35	40	45	50
4	9	14	19	24	29	34	39	44	49
3	8	13	18	23	28	33	38	43	48
2	7	12	17	22	27	32	37	42	47
1	6	11	16	21	26	31	36	41	46

55	54	53	52	51
60	59	58	57	56
65	64	63	62	61
70	69	68	67	66
75	74	73	72	71
80	79	78	77	76
85	84	83	82	81
90	89	88	87	86
95	94	93	92	91
100	99	98	97	96

Using *Hopping Hundred* with a Group

Playing *Hopping Hundred* requires people to multiply and divide. It also encourages them to think strategically.

About Multiples and Factors

Before your group plays *Hopping Hundred*, make sure everyone understands what multiples and factors are.

A *multiple* is what results when you multiply a number by other numbers. Some multiples of the number 3, for instance, are 6 (which is 3×2), 15 (which is 3×5), and 33 (which is 3×11).

Factors are numbers you can multiply together to get the number you're after. Some factors of the number 90, for example, are 2, 3, and 5, because $2 \times 3 \times 3 \times 5 = 90$. Other factors of 90 are 6, 9, 10, 15, 18, 30, and 45. All these numbers divide evenly into 90, leaving no remainder.

To find out if members of your group know what multiples and factors are, give them a number and ask for some multiples or factors of that number. Figuring out multiples and factors is extremely important in understanding and using fractions. (For more on that, see "Where's the Math?" on page 60.)

Playing *Hopping Hundred*

Introducing the Game

You can summarize the rules of the game aloud or have players follow the copied instructions.

Strategy in *Hopping Hundred*

This game requires players to think ahead. Playing smart means thinking not only about the number you are going to choose but also about the number your opponent might choose—or will be forced to choose—when it's his or her turn. A player wins by picking a number that has no multiples or factors left on the table.

When everyone has played a few games, discuss the strategy behind the game. Did anyone find a consistent technique for winning?

If no one has figured out a winning strategy, mention that there is a secret to winning—and suggest they play a few more games.

The Winning Secret: Playing with Primes

Many games of *Hopping Hundred* end with one player choosing a prime number greater than 50. A *prime number* is a number that can be evenly divided only by itself and 1. Because the prime number chosen is greater than 50, it has no multiples on the game board. Its only factor (other than itself) is 1.

So, the player's opponent is forced to choose 1, leaving the first player free to choose another prime number over 50—and leaving the opponent with nowhere to hop!

If your group has made *Eratosthenes' Sieve*, members will be familiar with prime numbers—and they will probably figure out the winning secret.

After people play this game a few times, it should become obvious why the rules require the first player to start with an even number. Otherwise, the first player could start with a large prime number and win the game very quickly!

Another challenge you could give your group is to try to make the game last as long as possible. If the two players work together, how many numbers can they fill up before they get stuck on a large prime number?

Where's the Math?

Playing this game will help people gain an understanding of multiples and factors. This, in turn, will help them when they are working with fractions, something that gives many people trouble.

As part of a problem involving fractions, a middle schooler may be asked to change one fraction into another fraction that is equivalent, or equal, to the original. For instance, a problem might ask for a fraction equivalent to $\frac{1}{15}$.

One way to find an equivalent fraction is to multiply the numerator and denominator by the same number. (When you do this, you are really multiplying the fraction by 1, because a number divided by itself is equal to 1.)

For example, $\frac{1}{15}$ can be changed into an equivalent fraction by multiplying it by $\frac{2}{2}$:

$$\frac{1}{15} \times \frac{2}{2} = \frac{2}{30}$$

Creating equivalent fractions comes in handy when a problem calls for adding or subtracting fractions. For example, consider this problem:

$$\frac{5}{15} + \frac{7}{45} = ?$$

The first step in adding the fractions is to change each fraction into an equivalent fraction—and then choose the equivalent fractions so that both have the same denominator.

An astute *Hopping Hundred* player will quickly realize that 45 is a multiple of 15 because $3 \times 15 = 45$. Now, you can make $\frac{5}{15}$ into an equivalent fraction with 45 as its denominator by multiplying it by $\frac{3}{3}$:

$$\frac{5}{15} \times \frac{3}{3} = \frac{15}{45}$$

And suddenly the addition problem looks much simpler:

$$\frac{15}{45} + \frac{7}{45} = \frac{22}{45}$$

If people in your group are having trouble with fractions, a few games of *Hopping Hundred* may be the first step to helping them out!

Tic-Tac-Toe Times

This game is fun to play, and the rules are simple. While members of your group are playing *Tic-Tac-Toe Times*, they are also reviewing the multiplication table. Playing strategically will be easiest for those who already know the multiplication table fairly well.

Preparation and Materials

Each pair of players will need:

- 18 tokens of one color and 18 tokens of another color (for example, scraps of paper of two colors, dried beans of two varieties, or 18 pennies and 18 nickels)

- 2 small objects (such as paper clips) to use as pointers

- copies of *Playing Tic-Tac-Toe Times* and *Tic-Tac-Toe Times Game Board*

- a copy of *Advanced Tic-Tac-Toe Times Game Board* (optional)

Using This Activity

Tips for how to lead a group in playing this game begin on page 66. *Tic-Tac-Toe Times* can also be played independently by groups of two—simply give them copies of the instructions and the necessary playing pieces.

Most players will find enough challenge in playing with the *Tic-Tac-Toe Times Game Board*, which requires them to remember the multiplication table up to 9×9. If some want to try a more challenging version of the game, give them the *Advanced Tic-Tac-Toe Times Game Board*, which requires players to remember the multiplication table up to 12×12.

Before playing this game, you might want your group to do *Eratosthenes' Sieve*. This will help anyone who has difficulty with the multiplication table, which will make playing *Tic-Tac-Toe Times* more fun.

Planning chart

Playing *Tic-Tac-Toe Times*	20 minutes

Playing Tic-Tac-Toe Times

Can you be the first to get four squares in a row?

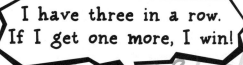

I have three in a row.
If I get one more, I win!

What Do I Need?

◇ 18 tokens of one color and 18 tokens of a different color

◇ two paper clips to use as pointers

◇ a copy of the *Tic-Tac-Toe Times Game Board*.

Here's How to Play

1 Player 1 places the two pointers on any numbers in the Factors Row. Player 1 multiplies these numbers together and places a token on the number he or she gets in the Products Box. (When you multiply numbers together, the number you get is the *product*. The numbers you multiply are *factors* of that product.)

2 Player 2 moves *one* of the pointers to a new number in the Factors Row. (The first move of the game is the only one in which both pointers are moved.) Player 2 multiplies the two marked numbers in the Factors Row to make a new product and puts a token on that product. If that product is already taken, Player 2 must choose a different number in the Factors Row.

3 Players keep taking turns. With each turn, a player moves *one* of the pointers, multiplies the two numbers in the Factors Row to get the product, and then covers that product in the Products Box. (If there's no move a player can make, that player loses the game!)

4 Both pointers can be placed on the same number. For example, both pointers could be placed on 5. The product, 5 x 5, would be 25.

5 The first person to cover four products in a row with no spaces between the products is the winner. The row can be horizontal, vertical, or diagonal.

6 If your opponent makes a mistake in multiplication and you spot it, you can capture a square in the Products Box for yourself. Name the correct product of the two factors in the Factors Row, and capture that box with one of your tokens.

Tic-Tac-Toe Times Game Board

Products Box

1	2	3	4	5	6
7	8	9	10	12	14
15	16	18	20	21	24
25	27	28	30	32	35
36	40	42	45	48	49
54	56	63	64	72	81

Factors Row

1	2	3	4	5	6	7	8	9

The Math Explorer
Published by Key Curriculum Press / © 2003 Exploratorium

Advanced Tic-Tac-Toe Times Game Board

Products Box

11	12	14	15	16	18	20
21	22	24	25	27	28	30
32	33	35	36	40	42	44
45	48	49	50	54	55	56
60	63	64	66	70	72	77
80	81	84	88	90	96	99
100	108	110	120	121	132	144

Factors Row

1	2	3	4	5	6	7	8	9	10	11	12

Playing *Tic-Tac-Toe Times* with Your Group

This game is a fun way to review and practice the multiplication table. If you want to make sure that all members of your group know the multiplication table before they play this game, we suggest you precede *Tic-Tac-Toe Times* with *Eratosthenes' Sieve*, an activity that helps people find patterns in the multiplication table.

Playing *Tic-Tac-Toe Times* also encourages people to think strategically.

Introducing *Tic-Tac-Toe Times*

Summarize the rules of the game aloud (page 62), or have players follow copies of the instructions. Once they understand the rules, they can play independently in pairs.

Each pair will need a copy of *Playing Tic-Tac-Toe Times* and *Tic-Tac-Toe Times Game Board*. Each pair will also need tokens and pointers.

When you go over the rules, be sure to mention that a player can put both pointers on the same factor to multiply a number by itself. Explain that the winning four in a row can be vertical, horizontal, or diagonal. The winning four must be right next to each other, with no spaces between them.

If all the products that a player can make are already taken, then there is no move the player can make, and he or she loses the game.

Where's the Math?

Playing this game not only helps reinforce players' knowledge of factors and products but also helps people who are struggling with fractions. Specifically, understanding multiples and factors can help them add, subtract, and simplify fractions. For more on this idea, see *Eratosthenes' Sieve* (page 46) and *Hopping Hundred* (page 55).

Finding a Winning Strategy

Players will eventually discover that there is more to this game than simply getting their own tokens on four squares in a row. They must also think about how to block their opponents' attempts to get four in a row.

As a game goes on, a player may have to strategically avoid certain factors to prevent his or her opponent from getting a critical winning square.

Making the Game More Challenging

The player who goes first can gain an advantage by taking one of the center four squares (18, 20, 28, or 30) on the first move. If players using the regular *Tic-Tac-Toe Times Game Board* are evenly matched, this advantage can mean that the player who goes first will win.

If your group notices this, we suggest you add one more rule: the player who goes first can't use the center four squares. This change in rules will force group members to come up with new strategies for winning.

Players who want to play a more challenging version of the game can use the *Advanced Tic-Tac-Toe Times Game Board*.

Tricks and Puzzles

Card tricks are mysterious and magical. It's fun—and challenging—to try to figure out how a card trick works.

In this section, you'll find two card tricks that look like magic and teach about math. You'll also find a tricky challenge that most people can't resist trying.

Magic Grid

This card trick teaches skills related to grids and graphing, topics covered in middle school math classes.

Preparation and Materials

To play as a group, you will need:
- 1 deck of cards for every 3 pairs of players

If you choose to have players learn the trick on their own, you will also need:
- 1 copy of *Magic Grid* and *How Does Magic Grid Work?* for each pair of players

Try the trick at least once yourself before using it with your group. You should also read *How Does Magic Grid Work?* so you can help your group understand the trick.

Using This Activity

You can teach your group this trick by demonstrating it and then having people do the steps along with you as you do it again. When Eva Jo Meyers of the Boys and Girls Club of San Francisco tried this trick with her group, she stressed performance. "At first we practiced laying out the cards in a 'magicianlike way,' then collected them suavely."

If members of your group are good at following written instructions, you might have them read the instructions and learn the trick on their own. If your group has a "homework hour," learning a card trick independently can be a great activity for those who finish their homework early. If you want people to learn the trick independently, give each person or pair a deck of cards and copies of *Magic Grid* and *How Does Magic Grid Work?* They can prove they've learned the trick by showing it to the rest of the group.

More tips for how to use *Magic Grid* start on page 75.

Planning chart

Demonstrating the trick	5 minutes
Having individuals try the trick	10 minutes
Figuring out how the trick works	10 minutes

Magic Grid

When you do this card trick, people may think you can read minds! You'll lay a bunch of cards on the table and have a friend choose one. With a little help, you'll use the power of math to figure out which card your friend chose.

What Do I Need?

◇ at least 16 playing cards

◇ a partner

What Do I Do?

Step 1 Deal 16 cards into a 4-by-4 grid, something like this:

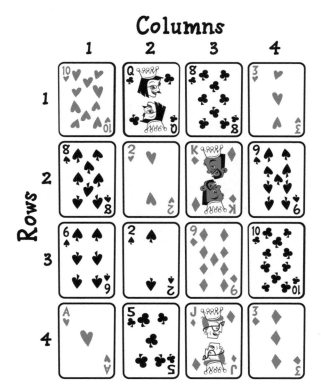

Step 2 Ask your friend to pick a card without touching it or telling you which it is.

Step 3 Ask your friend which column his or her card is in: Column 1, Column 2, Column 3, or Column 4? Memorize the number your friend tells you.

Step 4 Pick up all the cards you just dealt—but in a special way. Gather up the four cards in Column 1. (You don't have to keep them in any special order; just make sure all the cards in the first column are together.) Put the stack *face up* in your left hand. Then gather the cards in Column 2 and put this stack *face up* in your left hand, *on top* of the first stack. Do this for Column 3 and Column 4 as well.

Step 5 Flip the stack of cards over, and deal out four *rows* of four cards, face up. Start with the top row, which will be Row 1. Then deal out each row below the previous row. Be sure to lay the cards out in rows (running from left to right), not in columns (running up and down).

In the illustration, you can see that the cards that were in *Column 2* are now in *Row 2*.

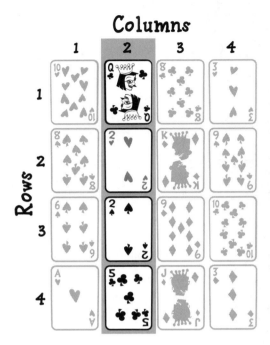

Let's say your friend tells you the card is in Column 2.

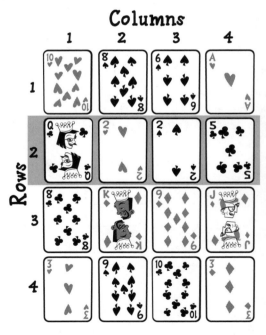

Redeal the cards. Now the second *column* from the old grid is the second *row* in this grid.

Step 6 Ask your friend which column the card is in now: Column 1, 2, 3, or 4?

Step 7 Here's the part that looks like magic. Your friend has told you two numbers. Count down to find the *row* that goes with the first number your friend gave. Then count over from the left to the *column* that matches the second number your friend gave. This is the card your friend chose.

Step 8 Now that you know which card your friend chose, you can play it up by putting your fingers to your temples and making some mysterious, magical sounds of deep concentration. Say something like, "The oracle tells me the card you chose is . . ." Then point to your friend's card. Your friend will know you did something sneaky but probably won't be able to figure out exactly *how* you did it.

Can you figure out how this trick works?

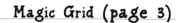
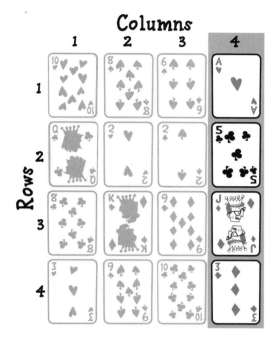

Suppose your friend says the card is now in Column 4.

If your friend said "Column 2" the first time and "Column 4" the second time, you'd find the card where Row 2 and Column 4 meet. In this example, that's the five of clubs, your friend's secret card.

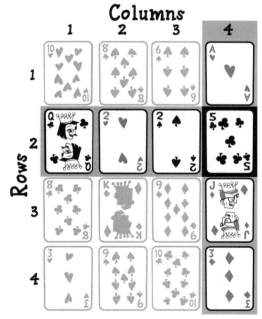

How Does Magic Grid Work?

This trick relies on the mathematical magic of grids.

What's a Grid?

You've probably seen grids before. A grid is a pattern of crossing horizontal and vertical lines. The vertical and horizontal lines form lots of little boxes.

A grid can help you find an exact location using only two numbers. In this card trick, you use the two numbers your friend gives you to find the location of the card your friend chose.

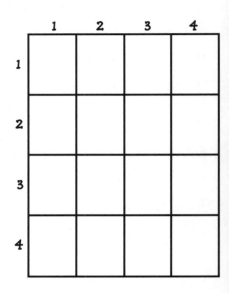

How Does That Work?

Imagine that there's a grid on top of the cards you laid out. It would look something like the picture at right:

If someone tells you which column and row a card is in, you can identify the card—simply by finding the column and the row. Where the row and column intersect, there's the card. If your friend tells you the card is in Column 3 and Row 2, you can find the king of diamonds—no problem! But that's not much of a trick.

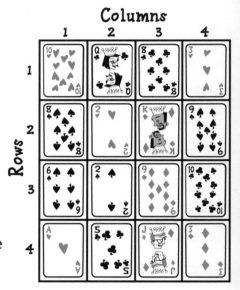

Here's what makes the *Magic Grid* trick so tricky. When you pick up the cards and deal them out again, you turn the columns of the first grid into the rows of the second grid. All the cards that were in Column 1 are now in Row 1. To see that this is true, check the illustrations on page 72 (or get out your cards and try the trick again).

When you ask your friend the second time which column the card is in, you already know which row it's in. (Remember, your friend told you which column the card was in. That column is now a row—and you know the card is somewhere in that row.) When your friend tells you the column the second time, you know which card in that row is the right card.

The Math Explorer
Published by Key Curriculum Press / © 2003 Exploratorium

Using *Magic Grid* with a Group

These instructions will help you teach your group the *Magic Grid* card trick. After you have amazed your group with your magical abilities, everyone will want to learn the trick.

Learning the Trick

Try the trick on your own, following the steps in *Magic Grid* (page 71). Then demonstrate the trick to your group, asking someone to select a card by pointing to it while you turn your back.

You can demonstrate the trick again while members of the group follow along. Or, if you want pairs to work independently, give each a copy of *Magic Grid.* Each pair will need at least 16 cards.

Have each person try the trick on his or her partner.

Figuring Out the Trick Together

After everyone has mastered the trick, ask people if they can determine how it works. Some may be able to figure it out without assistance, but most will need some help.

How Does Magic Grid Work? on page 74 will help your group figure it out. This explanation is written so that most members of your group can read and understand it, but we suggest that you lead the group in discovering how the trick works rather than handing out an explanation.

Here's one way to help everyone figure out how the trick works. Set up the grid of cards, and have someone choose a card while your back is turned. This time, ask which

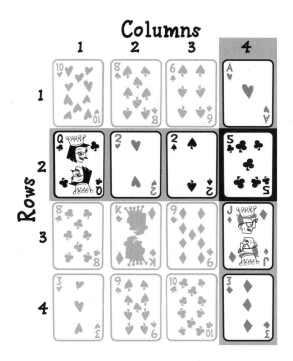

column and row the card is in. With this information, it's easy for you to identify the card—and the group will realize that.

Shuffle the cards and lay them out again. Repeat the original trick—but this time have four people each memorize one of the four cards in the column containing the selected card.

Gather the cards and lay them out again. Now ask the four people to find their cards. Point out which row these cards are in. This will demonstrate that the columns of the first grid have become the rows of the second grid. Finish the trick by asking which column the selected card is in.

Going Further

If some members of your group like this trick, ask them if they can figure out how to do the same trick with 25 cards in a 5-by-5 grid, or 36 cards in a 6-by-6 grid. The trick will work no matter what size the grid is, as long as it's a square grid with the same number of columns and rows.

Other Places to Find Grids

Battleship

If members of your group are familiar with the game of *Battleship*, they know how to use a grid. In *Battleship*, players use two numbers (or a letter and a number) to locate "battleships" on a grid.

Grids on Maps

Many maps are divided into grids to make finding a particular place easier. Across the top of the map, the columns of the grid are usually labeled with numbers. Along the side, the rows are usually labeled with letters.

Suppose you're trying to find a certain park on a map. You can look up the name of the park in the map's index. According to the index, the park is at "B3." Now all you have to do is find the row labeled "B" and the column labeled "3." Locate the grid square in this row and column, and that's where you'll find the park!

Grids in Graphs

You also see grids in many graphs. In most graphs, the grid lines are numbered, rather than the columns and rows. But the idea is the same: each point on the graph can be located using only two numbers.

In math classes, middle school students are introduced to graphing on a grid formed by a horizontal number line and a vertical number line.

Where's the Math?

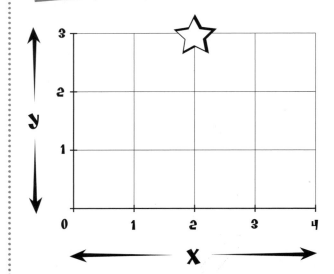

Any point on this grid can be located using two numbers: the number on the horizontal number line (also called the *x*-axis) and the number on the vertical number line (also called the *y*-axis). The point marked with a star, for instance, is located at 2 on the *x*-axis and 3 on the *y*-axis. Another way of stating this is to say that its *x*-coordinate is 2 and its *y*-coordinate is 3. (*Coordinate* is a mathematical word for location.)

Together, the *x*- and *y*-coordinates form an ordered pair, written as (x, y). It's called an *ordered pair* because the order is very important: if the order gets scrambled, you won't find the same location. The star's location can be written as the ordered pair (2, 3).

The Math Explorer
Published by Key Curriculum Press / © 2003 Exploratorium

Mind Reader

When you show this card trick to your group, everyone will want to learn how to do it. Once people learn the trick, they are very motivated to figure out how it works!

Preparation and Materials

For each pair of players, you will need:

- a deck of cards
- a copy of *Mind Reader*
- a copy of *Analyzing Mind Reader* (optional)
- a pencil and paper

Using This Activity

Instructions for how to lead a group through this card trick start on page 83. Try the trick at least once yourself before using it with your group. You should also read *Analyzing Mind Reader* so you can help your group understand the "magic" behind this trick.

The trick can also be learned independently by individuals or pairs. It's a great activity for those who finish other assignments early. Simply give each person or pair a deck of cards, a copy of the instructions, and the optional analysis, which explains why the trick works. You can ask those working independently to demonstrate that they've learned the trick by trying it on you.

Though the analysis is meant for those learning the trick independently, it can also be used as a reference to help you understand and explain the trick to groups. People who haven't been introduced to algebra may have trouble understanding the explanation of the trick—but they'll still have a great time learning it!

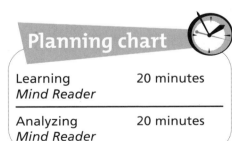

Planning chart

Learning *Mind Reader*	20 minutes
Analyzing *Mind Reader*	20 minutes

Mind Reader

These instructions tell you how to do a card trick that's guaranteed to make you look like a mind reader. Try it with a friend, and then spring it on some unsuspecting victims—like your family!

What Do I Need?

◇ a deck of cards

◇ a partner

◇ a pencil and paper

What Do I Do?

Step 1 Remove the tens and the face cards (jacks, queens, and kings) from the deck, and set them aside. Shuffle the cards.

Step 2 Ask your friend to pick a card from the deck. Tell your friend not to show it to you, but to memorize the number on it.

Step 3 Have your friend put his or her card on the table, face down.

Step 4 Take a card from the deck. *Don't show your card to your friend*.

Step 5 Memorize the number on your card, and then lay the card face down on the table next to your friend's card. Be sure your card is to the *right* of your friend's card *from your friend's point of view*. (If you and your friend are standing on different sides of the table, your card will be on the left from your point of view.)

From your friend's point of view

Step 6 Give your friend the paper and pencil so that he or she can do some math. Tell your friend to do the following things—*without showing you the answers.*

Here's what you tell your friend to do:	Suppose your friend chooses a 3:
Double the value of your card.	**3 × 2 = 6**
Add 2 to that number.	**6 + 2 = 8**
Multiply the result by 5.	**8 × 5 = 40**

Step 7 It's your turn to do a little math in your head.

Here's what you do:	Suppose your number is 8:
Subtract the number of your card from 10.	**10 − 8 = 2**

Step 8 Now it's your friend's turn again.

Here's what you tell your friend to do:

Tell your friend your answer from Step 7. Have your friend subtract that from the number he or she got after doing all the math in Step 6 and then tell you the final result.

40 − 2 = 38
Your friend gets 38.

Step 9 Here's the magic part. Flip over your friend's card and then your own. There will be the two digits of the number your friend just gave you. Enjoy the shocked look on your friend's face!

Step 10 But wait! Suppose your friend came up with a different number. Mutter something like, "I know the magic works—so let's check your math." Check the math—and you will get the number that's on the cards.

Your friend's card   Your card

Analyzing Mind Reader

Try this trick a few times, and you'll soon realize that it doesn't matter what numbers you or your friend choose. The trick works no matter what.

What's Going On?

When you lay out the two cards side by side, the numbers on them form a two-digit number. In a two-digit number, one digit tells you how many 1s there are—that's called the *ones place*. The other digit tells you how many 10s there are; that's the *tens place*. The number 38 has a 3 in the tens place and an 8 in the ones place. It has 3 tens and 8 ones.

All the arithmetic your friend did was merely a roundabout way of moving his or her number to the tens place and putting your number in the ones place.

A Simple Version

To see what's going on, let's do a simpler version of the trick. Suppose you tell your friend to multiply his or her number by 10. If your friend's number is 3, then

$$3 \times 10 = 30$$

and his or her number has moved from the ones place to the tens place.

Now suppose you tell your friend your number and have him or her add that to the result. If your number was 8, then

$$30 + 8 = 38$$

When you turn over the cards, you will reveal the number 38—but your friend might not be impressed. He or she might be able to figure out what you did.

Getting Tricky

The *Mind Reader* instructions give a tricky way of hiding what you are doing from your friend.

To see what's going on, let's go through the steps one at a time. Because it doesn't matter what number your friend chooses, we'll simply represent the number with the symbol □. This square could be any number from 1 to 9.

First, you told your friend to double his or her number. That's his or her number times 2. You can write this mathematically like this:

$$\square \times 2$$

Next you told your friend to add 2:

$$(\square \times 2) + 2$$

Then you told your friend to multiply this by 5:

$$((\square \times 2) + 2) \times 5$$

This looks pretty complicated, but what you are really doing is adding two numbers—(□ × 2) and 2—and multiplying the result by 5. That's the same as multiplying each number by 5 and then adding them together.

This: $((\square \times 2) + 2) \times 5$

is the same as this: $((\square \times 2) \times 5) + (2 \times 5)$

is the same as this: $(\square \times 10) + 10$

In Step 8 of the trick, you had your friend subtract the difference between 10 and your number. Suppose your number was 1. You will ask your friend to subtract 9, which is the difference between 10 and 1:

$$(\square \times 10) + 10 - (10 - 1)$$

$$(\square \times 10) + 10 - (9)$$

$$(\square \times 10) + 1$$

What you end up with is your friend's number times 10, plus your number—the same numbers that are on the cards when you flip them over!

Getting Even Trickier

The instructions for *Mind Reader* give one way to move a number from the ones place to the tens place. Can you come up with another set of tricky instructions for doing the same thing? If so, you can make up your own version of this trick.

Algebra—The Great Unknown

When numbers in math problems are represented by symbols, it's called *algebra*. What's great about algebra is that it lets you find answers to math problems even when you don't know all the numbers involved. (In this case, your friend's number and your number could be anything from 1 to 9.) Usually, unknown numbers in algebra problems are represented with letters, not squares. But it works just the same.

   X is the letter most commonly used to represent an unknown quantity. Have you ever heard of the TV show *The X-Files*? It's a show about weird, unexplained things like alien creatures and people with strange powers. Just as in algebra, the *X* in *The X-Files* stands for the unknown.

The Math Explorer
Published by Key Curriculum Press / © 2003 Exploratorium

Using *Mind Reader* with a Group

These instructions will help you teach your group the *Mind Reader* card trick. This trick requires no manual dexterity (though a little mental dexterity helps).

Demonstrating the Trick

We suggest you read *Mind Reader* and try the trick a few times before showing it to your group. Once you are confident you can do it, demonstrate the trick to your group.

First, ask everyone to get out a pencil and a sheet of paper to do some simple arithmetic. Turn your back while the group chooses a card and places it face down on a table or desk. Then choose a card yourself, memorize it, and put it down just to the right (from the group's perspective) of the card the group has chosen.

Then follow the steps in the *Mind Reader* instructions. Have group members follow your directions. When they have reached the end of their calculations, turn over the cards and reveal the number they have calculated.

Your group will be amazed!

Having Everyone Try the Trick

Have the group divide into pairs, and hand out decks of cards and copies of the *Mind Reader*. Have each pair do the trick.

You may need to point out that the magician doesn't actually know the number until the cards are turned over. The magician only knows the card he or she chose. Turning over the cards reveals the number that is the inevitable result of the other person's calculations. If he or she came up with a different number, the magician and that person should check the math together. If the math is done correctly, the number will match the cards.

Figuring Out How the Trick Works

Analyzing Mind Reader explains how the trick works. If your group can understand "A Simple Version," explain that the trick is doing the same thing, but with lots of extra steps. Basically, Step 6 multiplies the friend's number by 10 and adds 10 to the result. Then Steps 7 and 8 subtract the magician's number from 10 and subtract the difference from the result of Step 6. The value of the magician's card is, in effect, added to 10 times the friend's number.

Younger middle schoolers may not be able to follow the entire explanation, but they may see that the card the victim chooses always moves over to the tens place. Even if they don't understand the complete explanation, middle schoolers of all ages enjoy doing this trick.

Where's the Math?

Learning and practicing this trick involves basic arithmetic skills. Figuring out this trick requires problem solving skills and makes people think about *place value* (how the value of each digit in a number depends on its position). Understanding the trick requires an understanding of some basic algebra and the *distributive property of multiplication*—$a(b + c) = ab + ac$.

Exponential Folding

Can you fold a sheet of newspaper in half 10 times? Newspaper is relatively thin, and a sheet of it is rather large, so most people mistakenly assume they could easily accomplish this. This simple and surprising activity offers an opportunity to introduce your group to exponents and exponential growth.

Can you fold a sheet of newspaper in half 10 times?

Preparation and Materials

For each person, you will need:

- a copy of *Exponential Folding*
- a sheet of newspaper (the larger, double-page size)
- a calculator (optional)

Using This Activity

This quick activity provides an interesting challenge for your group. As an incentive, you may want to offer a prize to anyone who can fold a newspaper in half 10 times. (We think it's impossible! The most anyone we know has managed is 8 folds.) Once the group discovers that 10 folds is impossible, you may choose to offer the prize to whoever gets the most folds—and you may have multiple winners.

Tips for using *Exponential Folding* start on page 86.

Planning chart	
Folding the newspaper	15 minutes
Visualizing exponential growth	15 minutes

Exponential Folding

Do you think you can fold a large sheet of newspaper in half 10 times?

Can you fold a sheet of newspaper in half 10 times?

What Do I Need?

◇ a sheet of newspaper (the larger, double-page size)

◇ a calculator (optional)

What Do I Do?

Step 1 Fold your sheet of newspaper in half. It doesn't matter which direction you fold it, as long as you fold it in half. How many layers of newspaper do you have after folding it in half once? You should have two layers.

Step 2 Without unfolding your first fold, fold the newspaper in half again. Now you have two folds. Only eight more to go. How many layers of newspaper do you have after making the first two folds? (You should have four layers.)

Step 3 Fold the newspaper in half again. The area of the folded paper will be getting smaller, and the thickness of the stack of folds will be increasing. How many layers of newspaper do you have now? Use the chart to keep track.

Step 4 Fold the newspaper in half again. And again. Keep track of the number of folds you have made as you go along. Keep track of the number of layers of newspaper, too.

Step 5 At some point, you will find that you cannot make the next fold. The area of the folded paper is too small, and there are too many layers. How many folds were you able to complete? How many layers of newspaper do you have?

Number of Folds	Number of Layers of Newspaper
0	1
1	2
2	4
3	8
4	
5	
6	
7	
8	
9	
10	

Exploring Exponents with *Exponential Folding*

This tricky challenge can lead to some interesting discussion.

Some Questions to Ask Your Group

Ask people how many times they folded the paper. Then ask why they couldn't fold it any more. They'll probably say something like, "It got too thick." Here are some questions to ask to help them gain a mathematical understanding of why they couldn't fold the paper 10 times: How many layers of paper are there after one fold? After two folds? After three folds? After four folds?

Looking for Patterns

If people didn't pay attention to how many layers of paper they had after each fold, have them unfold their newspapers and refold them, counting the layers after completing each fold.

- After 1 fold, they'll have 2 layers of paper.
- After 2 folds, they'll have 4 layers of paper.
- After 3 folds, they'll have 8 layers of paper.
- After 4 folds, they'll have 16 layers of paper.

Do they see a pattern here?

To help them find a pattern, write what they know so far on the board or somewhere that the whole group can see. Organize the information into a table like the one here. Don't fill in all the numbers in the second column.

Number of Folds	Number of Layers of Newspaper
0	1
1	2
2	4
3	8
4	
5	
6	
7	
8	
9	
10	

Ask your group to look for a pattern. Let members provide you with the "Number of Layers of Newspaper" data as you fill in the table. Ask if anyone can describe the pattern. One way is to say that each time the paper is folded, the number of layers doubles. The completed table should look like this:

Number of Folds	Number of Layers of Newspaper
0	1
1	2
2	4
3	8
4	16
5	32
6	64
7	128
8	256
9	512
10	1024

Now people can see why folding the paper gets so difficult. After the eighth fold, they are attempting to fold 256 sheets of paper—and that's hard.

Where's the Math?

Each time the paper is folded, the number of layers doubles. We can find the number of layers by multiplying the previous number of layers by 2. Another way to describe this is to say that the number increases by a factor of 2. After one fold, there are 2 layers. After two folds, there are 2×2, or 4, layers. After three folds, there are $2 \times 2 \times 2$, or 8, layers.

A shorthand way to write $2 \times 2 \times 2$ is 2^3. The 3 is an *exponent*, and the 2 is the *base*. The exponent says how many times the base is used as a factor. So, 2^5 means $2 \times 2 \times 2 \times 2 \times 2$, or 32. The expression 2^2 can be read as "two squared" or "two to the second power." The expression 2^3 can be read as "two cubed" or "two to the third power." The expression 2^5 can be read as "two to the fifth power."

Thinking about exponents can make it easier to keep track of the number of layers of paper. After two folds, there are 2^2, or 4, layers. After three folds, there are 2^3, or 8, layers. After 10 folds, there would be 2^{10}, or 1024, layers. Notice how the exponent is the same as the number of folds. So, what does it mean when there are zero folds? When there are zero folds, there is one layer, and 2^0 is equal to 1. Any number to the zero power is equal to 1.

Exponents and Population

A mathematician would say that the number of layers of paper grows exponentially. Ecologists and biologists think about exponential growth in relation to populations.

A population that isn't kept in check by disease or predators may grow exponentially. In studying populations, scientists talk about a population's *doubling time*—that is, how long it takes for it to double in size. A population that grows at a rate of 5 percent will double in size in 15 years.

As your experience with folding paper has shown, doubling an amount will get you to large numbers very fast! That's one reason people worry about population growth.

A classic exponent problem that you may want to share with your group involves a fictitious population of bacteria. These bacteria live in a closed bottle. The population of the bacteria doubles every minute. At 12 o'clock midnight, the bottle is completely full. What time was it when the bottle was half full? (11:59 P.M.) What time was it when the bottle was one quarter full? (11:58 P.M.) If you were one of the bacteria in the bottle, at what time would you start to worry that you were running out of space? (Answers will vary, but at 11:59, with only half the space occupied, occupants would be unlikely to perceive much of a space problem.)

Suppose some explorers from the bacteria population in the full bottle discover three empty bottles to help solve their population problem. How long will it take the population to fill these new bottles? (2 minutes)

Visualizing Exponential Growth

You can extend this activity by having your group make graphs of the tables you made together. Put the number of folds along the horizontal axis, or *x*-axis. Put the number of layers of newspaper along the vertical axis, or *y*-axis. The resulting graph will curve more and more steeply upwards. A curve that gets steeper and steeper is a characteristic of exponential growth.

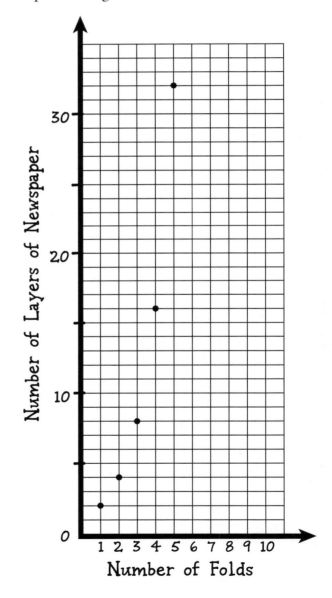

Cool Things to Make

Young people sometimes see math as abstract. In these activities, math becomes more concrete—and members of your group make something interesting to take home.

When your group makes boxes from greeting cards or draws large-scale versions of small cartoons, members may not be aware that they are exploring mathematical concepts. However, these activities will give them a hands-on feel for math that they might not get any other way.

Activity 12 **Colossal Cartoons**

While exploring a technique for enlarging small cartoons, people learn about scale, an important concept in middle school math.

Activity 13 **Greeting Card Boxes**

In the process of making beautiful boxes, people learn about area and volume.

Activity 14 **Jacob's Ladder**

The Jacob's ladder toy has fascinated people for centuries. In this activity, group members make their own toy and explore how it works.

Activity 15 **Paper Engineering**

Group members create amazing three-dimensional sculptures and pop-up greeting cards from ordinary paper.

Activity 16 **Incredible Shrinking Shapes**

This activity combines science and crafts. People shrink plastic recycled from deli containers to make medallions or earrings while learning about ratios and scaling.

ACTIVITY **12**

Colossal Cartoons

In *Colossal Cartoons*, people learn a technique they can use to draw a larger version of a small cartoon. In the process, they learn about scaling, a concept often discussed in middle school math classes. This activity also relates to graphing. Many mural artists use this technique to transform sketches into wall-size works of art.

Preparation and Materials

People can copy a cartoon from page 94, or they can bring in cartoons and draw their own favorite cartoon characters. Advise them that simple characters (those drawn with only a few lines) are easier to copy.

To copy the cartoons on page 94, each person will need:

*   a pencil
*   a ruler (They can share if they need to.)
*   copies of *Drawing a Colossal Cartoon*, *Gridded Cartoons*, and *Big Grid for Cartoons*

To copy their own cartoons, each person will also need:

*   a cartoon they don't mind drawing on
*   transparent tape
*   copies of *Preparing Your Cartoon* and *Big Grid for Cartoons*

Using This Activity

Tips for how to use *Colossal Cartoons* start on page 96.

Planning chart

Drawing a cartoon	about 45 minutes

The Math Explorer
Published by Key Curriculum Press / © 2003 Exploratorium

Drawing a Colossal Cartoon

Even if you're not a great artist, you can take a small picture and scale it up to make a big picture. You do this by copying one square at a time.

What Do I Need?

◈ a copy of *Gridded Cartoons* or a cartoon you have prepared yourself

◈ a copy of *Big Grid for Cartoons*

◈ a pencil

◈ a ruler

What Do I Do?

Step 1 The lines drawn on top of the cartoon make a pattern of squares known as a *grid*. Draw a rectangle that just fits around the cartoon, following the lines of the grid.

Step 2 Next to your cartoon, write down how many squares wide and how many squares tall your rectangle is.

Step 3 Write a letter beside each *row* of squares and a number under each *column* of squares.

You can use these numbers and letters to locate squares in the grid. For example, find Square B2 by finding Row B and then going over to Column 2.

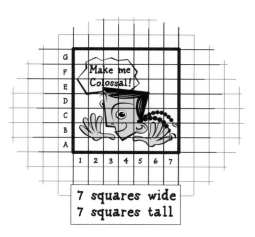

7 squares wide
7 squares tall

Step 4 In the middle of the *Big Grid for Cartoons,* draw a rectangle that's the same number of squares wide and tall as the rectangle you drew around your cartoon. This rectangle will be larger than the one around your cartoon, but it will contain the same number of squares.

Step 5 Number the columns and letter the rows. When you're done, you will have something that looks like a large version of the rectangle around your cartoon.

Step 6 Each square in the grid on your cartoon has a matching square on the big grid paper. Start with a square on the cartoon that doesn't have many lines in it. Look for the darkest or longest line in that square. Find one place where it touches a side of the square. Where does it touch? Where does it go from there? Copy that line into the matching square of your big grid paper.

Step 7 Do the same thing with another square. Repeat this until you've copied the most important lines from all the squares onto the big grid. You now have a line drawing of the cartoon. Then go back and fill in the details, square by square.

Step 8 Your drawing of the cartoon is a scale drawing of the original cartoon. A *scale drawing* looks just like the original, but it's a different size. To enlarge the small cartoon, you made each square twice as tall and twice as wide as the original. That means you used a *scale factor* of 2.

Your new drawing is twice as tall and twice as wide as the original. How much area does it cover in comparison with the original? If you are tempted to say it covers twice as much area, think again!

The Math Explorer

Published by Key Curriculum Press / © 2003 Exploratorium

Preparing Your Cartoon

You can draw a large version of a cartoon that already has a grid drawn on it—or you can start with a cartoon that you like and draw your own grid. Here's how.

What Do I Need?

◇ a cartoon (9-by-11 cm or smaller)

◇ centimeter grid paper

◇ a pencil or pen that makes a line dark enough to see on top of your cartoon

◇ a ruler

◇ transparent tape

What Do I Do?

Step 1 Tape your cartoon in the middle of the grid paper.

Step 2 Line the ruler up with a grid line on either side of the cartoon. Using your pencil, draw the line so that it runs right across the cartoon.

Step 3 Draw all the lines of the grid on your cartoon.

When you've done that, you are ready to draw a scaled-up version of your cartoon. Follow the instructions in *Drawing a Colossal Cartoon.*

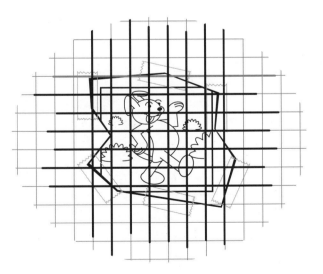

Gridded Cartoons

The Math Explorer
Published by Key Curriculum Press / © 2003 Exploratorium

Big Grid for Cartoons

Helping Your Group Make *Colossal Cartoons*

People in your group may have heard of "drawing to scale." A scale drawing looks just like the original, but it's a different size—either larger or smaller. In this activity, people change the scale of a cartoon to make a larger version.

Before Your Group Begins

Some students may not know the word *colossal*. You can tell them it is just a big word for "big."

Before your group can get started, everyone needs a cartoon with a grid drawn on it. Use the cartoons in *Gridded Cartoons* (page 94), or have people bring in their favorite cartoon characters and draw a grid on them by following the instructions on *Preparing Your Cartoon*. Advise them that simple characters (those drawn with only a few lines) are easier to draw. Snoopy from the *Peanuts* comic strip is much easier to draw than Spiderman, for example. You can bring in a few pages of the comics from the newspaper in case some of them forget. Be sure that the cartoons people choose are not larger than 9-by-11 centimeters, or the copied cartoon won't fit on the *Big Grid*.

Drawing a Colossal Cartoon

About Grids

Before people do any drawing, they need to put a letter beside each row of squares and a number under each column of squares. Once these are in place, people can locate a particular square if they know its letter and number. To find Square B2, for example, all they have to do is find the row labeled "B" and the column labeled "2," and identify the place where the row and column intercept. That's Square B2. Make sure everyone can correctly find a square in the grid. Ask everyone to find Square B2 (or C4 or D1)—and make sure they all locate the same square.

These pairs of letters and numbers are called *coordinates*. Make sure people can find a few different coordinates.

If members of your group play *Battleship*, they know how to find a location by using two coordinates. *Magic Grid*, the card trick on page 70, also uses a grid.

Some Drawing Tips

Drawing talent isn't required for this activity. You may have to tell your group this, because some people may freeze up and say, "But I can't draw!" What's required here is careful observation and careful copying.

If people ask for help, suggest that they find a square of the cartoon that doesn't have many lines in it. Then have them look for the darkest or most important line in that square and locate one place where this line touches a side of the square. Does it intersect with the middle of the side of the square? Near the top? Near the bottom? Where does it go from there? Suggest that they focus their attention on this one line and copy it carefully.

If they still have trouble, Joanne Roeper of the Community Housing Opportunity Corporation recommends that you have them cover all the boxes except the one they are working on.

Many people find it easier to copy the most important lines of the cartoon first, creating a simple line drawing. Once this line drawing is complete, they can go back and fill in the details and the shading.

About Scaling

Each square of the big grid is twice as tall and twice as wide as each square of the small grid. A mathematician would say that the scale factor in this activity is 2.

Step 8 of *Drawing a Colossal Cartoon* (page 92) asks a trick question: "Your new drawing is twice as tall and twice as wide as the original. How much area does it cover in comparison with the original?" (You may have to remind people that *area* measures the size of a surface. It's measured in square units, such as square centimeters or square inches.)

People may say the new drawing has twice the area of the original cartoon. If so, ask them how many copies of the small drawing they would need to completely cover the big drawing.

They can figure this out by putting the small grid on top of the big grid and holding them both up to a light. They can see that it takes four squares on the small grid to fill one square on the big grid. The area of the new drawing is thus four times the area of the small drawing.

The area of a rectangle can be calculated by multiplying the length of the rectangle by

its width. When the length and the width both double, the new area is four times the old area. (For more on this, see page 134 of *Incredible Shrinking Shapes*.)

This is important to keep in mind if your group wants to make a mural. Suppose members paint a small picture and then decide to make a version of that picture that is 10 times as tall and 10 times as wide. How much more paint will they need? They'll need 10×10, or 100, times as much paint, because the new picture will have 100 times the area of the original.

You might suggest that if they want to do this activity again, they can trace the big grid with pencil onto another sheet of paper and then erase the gridlines when they finish.

Where's the Math?

This activity can help people understand area, scaling, and scale factors, important topics in middle school math.

In math class, people may have learned about *similar figures*. These are figures that have exactly the same shape, though they may be different sizes. When people draw a large version of the rectangle that surrounds the small cartoon, they are drawing a similar figure.

This activity also provides an opportunity to practice skills related to graphing. In *Colossal Cartoons*, people use a number and a letter to identify a location on a grid. This is very similar to what they do when drawing a graph.

One way to draw a graph is to start with a horizontal line (usually called the *x-axis*) that crosses a vertical line (usually

called the *y-axis*). The place where the lines cross is called the *origin*.

The location of a point on a graph is determined by two numbers. The first number (the *x-coordinate*) tells how far to the right or left of the origin a point is. The second number (the *y-coordinate*) tells how far up or down from the origin the point is. Together, these two numbers determine exactly where to put the point.

In *Colossal Cartoons*, people use a number for the *x*-coordinate and a letter for the *y*-coordinate, rather than numbers for both. And they use these coordinates to identify a grid square, rather than a single point. But the principle is the same. Together, the number and letter specify an exact location on a grid.

The Math Explorer
Published by Key Curriculum Press / © 2003 Exploratorium

Greeting Card Boxes

In this activity, members of your group make boxes out of old (or new) greeting cards or postcards. But first, they practice making boxes from grid paper and, in the process, learn about area and volume. When you do this activity, every member of your group gets something nice to take home.

Preparation and Materials

For each member of your group, you will need:

- 1 rectangular greeting card or 2 postcards (Used cards are fine. You might ask people to bring in cards from home. Cards must be rectangular; square cards will not work for this activity.)

- a metric ruler (see *Centimeter Ruler*, page 202)

- copies of *Making a Greeting Card Box*, *Greeting Card Box Data Sheet*, and *Grid Paper* (page 204)

For each pair or small group, you will need:

- scissors

- pencils

- a calculator

- transparent tape (optional)

Using This Activity

Tips for using *Greeting Card Boxes* start on page 105. If you have time, make a box out of a greeting card before introducing this activity to your group. Displaying your completed box is a great way to encourage participation.

Planning chart

Making a box from grid paper and exploring area and volume	25 minutes
Making a box from a greeting card or postcards	15 minutes

Making a Greeting Card Box

Make a box from grid paper for practice—and then make a beautiful box from a greeting card.

What Do I Need?

◇ a sheet of centimeter grid paper

◇ a metric ruler

◇ a pencil

◇ scissors

◇ a calculator

◇ 1 greeting card or 2 postcards of the same size

◇ *Greeting Card Box Data Sheet*

What Do I Do?

Step 1 On the grid paper, draw a rectangle that has an area of 120 square centimeters, or 120 cm². To find the area of a rectangle, multiply the length by the width. So, you need two whole numbers that give 120 when you multiply them. When you multiply centimeters by centimeters, you get square centimeters, or cm².

Figure out two numbers that you can multiply together to get 120. These numbers will be the length and width of your rectangle.

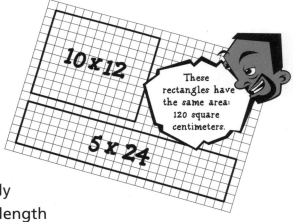

These rectangles have the same area: 120 square centimeters.

Step 2 Make sure the rectangle you want to draw will fit on your grid paper. If it won't, figure out two other numbers for your length and width.

Step 3 Draw your rectangle on the grid paper. Use your ruler and the grid lines to help you.

The Math Explorer
Published by Key Curriculum Press / © 2003 Exploratorium

Step 4 Write the length, width, and area of your rectangle on your *Greeting Card Box Data Sheet*.

Step 5 Cut out your rectangle.

Step 6 Turn the rectangle over to the blank side. Draw both diagonals on the blank side of the rectangle.

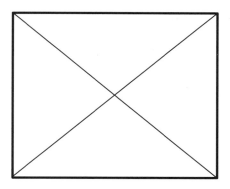

Step 7 The place where the diagonal lines cross is the center of your rectangle. Fold in the sides of your rectangle so that their edges meet in the center of the rectangle. Then unfold them.

Step 8 Fold the top and bottom sides of your rectangle so that their edges meet in the center. Then unfold them.

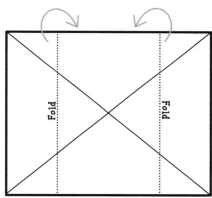

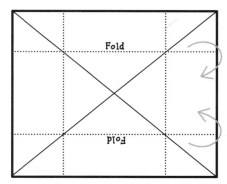

Step 9 Make four cuts into the rectangle. Cut into the sides that are shorter (the widths), and only cut to the point where the fold and the diagonal meet.

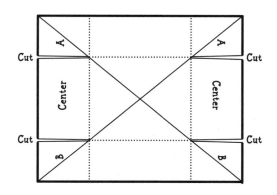

Step 10 Fold up Flaps A and B at each end of your rectangle, leaving the center flap between them lying flat. Then fold in both sides so that Flaps A and B meet or overlap.

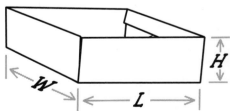

If you have a long, skinny rectangle, you may need to trim Flaps A and B to fit within the box sides you have just created.

Step 11 At each end, fold the center flap up and over Flaps A and B. The center flap will hold Flaps A and B in place.

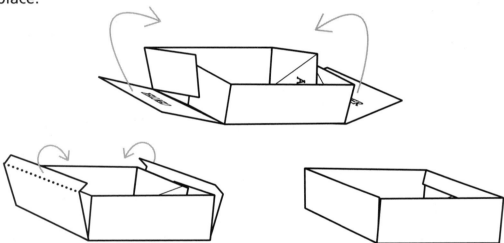

Step 12 Measure the length, width, and height of your box in centimeters. Write your measurements in the first row of your *Greeting Card Box Data Sheet*.

Step 13 The *volume* of a box is the amount of space it fills. You can calculate the volume of your box by multiplying its length by its width by its height. When you multiply centimeters by centimeters by centimeters, you get *cubic centimeters*, or cm^3.

Calculate the volume of your box. Write the volume on your data sheet.

The Math Explorer
Published by Key Curriculum Press / © 2003 Exploratorium

Step 14 Collect information on the boxes that other people made. Put the information on your data sheet. Notice that everyone started with rectangles of the same area, but the boxes have different shapes and volumes.

Step 15 Now you're ready to turn a greeting card or two postcards into a beautiful box.

If you are using a greeting card, cut it in half along the fold, making two rectangles with the same area. The side of the card with a picture on it will become the box top. The side of the card where you would write a message will become the box bottom.

If you are using postcards of the same size, you do not need to cut them. One postcard will become the box top, and one will become the box bottom.

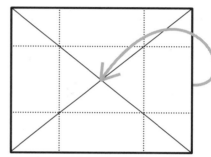

Step 16 Take the rectangle that has a picture on it, and turn it over so you are looking at the back side. (It's OK if it has writing on it. Just ignore that.) Follow Steps 6–11 to make the box top.

Step 17 Follow Steps 6–11 to make the box bottom from the other half of the card (or your other postcard). This time, fold the sides in a little beyond the center to make this box a little smaller than your first box. The box top will then fit on the box bottom, making a box with a lid.

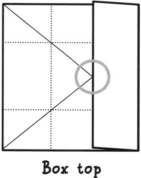

Box top

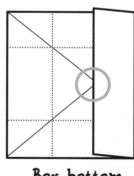

Box bottom

Step 18 Use your ruler to measure the length, width, and height of your box. Figure out its volume.

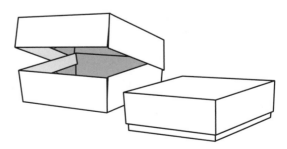

Greeting Card Box Data Sheet

Use this data sheet to record information about your box and the boxes that other people make.

Name of Box Maker	Length of Rectangle	Width of Rectangle	Area of Rectangle	Length of Box	Width of Box	Height of Box	Volume of Box
	cm	cm	cm²	cm	cm	cm	cm³
	cm	cm	cm²	cm	cm	cm	cm³
	cm	cm	cm²	cm	cm	cm	cm³
	cm	cm	cm²	cm	cm	cm	cm³
	cm	cm	cm²	cm	cm	cm	cm³
	cm	cm	cm²	cm	cm	cm	cm³
	cm	cm	cm²	cm	cm	cm	cm³
	cm	cm	cm²	cm	cm	cm	cm³
	cm	cm	cm²	cm	cm	cm	cm³
	cm	cm	cm²	cm	cm	cm	cm³
	cm	cm	cm²	cm	cm	cm	cm³
	cm	cm	cm²	cm	cm	cm	cm³
	cm	cm	cm²	cm	cm	cm	cm³

The Math Explorer
Published by Key Curriculum Press / © 2003 Exploratorium

Helping Your Group with *Greeting Card Boxes*

This activity will give your group some practice in making measurements in the metric system. Making boxes and comparing the volumes of different boxes will also increase understanding of area and volume, two important concepts in middle school mathematics.

Volume can be defined as the amount of space an object occupies. This can be difficult for people to visualize. Making these boxes transforms the abstract concept of volume into something concrete that people can get their hands on—and each member of the group ends up with a beautiful box.

You can choose to have people do this activity on their own, following the steps in *Making a Greeting Card Box*, or you can lead your group through the process step by step.

Making Boxes from Grid Paper

If you made a box from a greeting card, show it to your group at the beginning of the activity. These boxes are so pretty that most people will want one to take home. You may want to emphasize that doing this activity with the grid paper is practice for making a box out of the nicer card stock.

Drawing a Rectangle

The first thing people have to do is figure out how to draw a rectangle with an area of 120 cm^2. Because the area of a rectangle is the length multiplied by the width, they need two whole numbers that produce 120 when multiplied together. For example, a rectangle that is 20 cm long and 6 cm wide has an area of 120 cm^2.

If you use copies of the grid paper on page 204, you'll find that four different rectangles with whole number dimensions and an area of 120 cm2 will fit on the grid:

- 10 cm by 12 cm
- 8 cm by 15 cm
- 6 cm by 20 cm
- 5 cm by 24 cm

Encourage people to draw different rectangles, and remind them that everybody's rectangle will have the same area. Making a box from a rectangle that measures 5 cm by 24 cm is a little tricky. Make sure that whoever chooses to draw that rectangle is comfortable doing a little bit of tinkering to make the box work.

After people draw their rectangles, make sure they put the information about the dimensions in the first row of the table on the *Greeting Card Box Data Sheet*.

Drawing Diagonals

Next, people will find the center of their rectangles by drawing diagonal lines connecting the corners of the rectangles. Point out that they can find the center of any rectangle by drawing its diagonals. The intersection of the diagonals is always at the center of the rectangle.

Making the Box

You may want to demonstrate the rest of the steps in making a box. Simply follow the pictures in the instructions.

Here are a few tips:

- When folding the sides in, make sure the edges of the card meet right at the intersection of the two diagonals.

- The cuts are made into the sides that are shorter (the widths), and the cuts go only to the point where the folds and the diagonals meet.

- If people cut too far or cut into the wrong sides, they can fix their mistakes with tape.

Some people may need help folding the sides to make a box. People who are making boxes from long, skinny rectangles may have to trim Flaps A and B to make them fit when the box is folded. The center flap may just barely fold over Flaps A and B—or it may be long enough to fold over the flaps and reinforce the floor of the box. You may have to help people make these adjustments and reassure them that there are many different ways to make a box.

Finding the Volume

The volume of a box is its length times its width times its height. Have people measure their boxes in centimeters and record the measurements on their data sheets.

Some people may have trouble identifying the three different dimensions. For example, they may measure the length and the width, and then measure the length a second time, thinking it is the height. Remind them that most of the time, they will get three different measurements.

To get the volume, each person must multiply length by width by height. Tell them that this formula ($L \times W \times H$) gives us the amount of space the box fills in cubic centimeters. Have them record the volume on their data sheets.

Comparing Volumes

Ask the people in your group to collect information on the area and volume of different boxes from their peers, and record the information on their data sheets.

When everyone is done, ask the group what they notice. Perhaps they will see that the boxes have different shapes and volumes —even though everyone started with a rectangle of 120 cm^2. You may ask several people to hold up their boxes and share the areas and volumes they calculated.

Making Boxes from Greeting Cards or Postcards

Before your group begins *Making a Greeting Card Box*, make sure you have enough cards (new or used) so that everyone has at least one greeting card or two postcards that are the same size. Remember, square cards won't work.

Members of your group can simply follow the instructions to make the top of the box. Be sure they draw their diagonal lines on the back side of the cards, so that the pretty pictures will be on the outside of the boxes.

Making the bottom of the box is just the same as making the top—with one small change. For the bottom to fit inside the top, it needs to be a bit smaller than the top. To make a box that's smaller, simply fold the sides a little beyond the center in Steps 7 and 8. When the sides are folded up to make a box, the box will be a bit shorter and a little less wide—but a little taller than the box top. It should fit neatly inside the box top.

After everyone has finished, have them find the volumes of their boxes by multiplying the length by the width by the height. Ask volunteers to hold up their boxes and share the areas and volumes they calculated.

Where's the Math?

This activity gives people an opportunity to make metric measurements and practice finding factors. *Factors* are numbers that multiply together to make another number. The numbers 8, 10, 12, and 15 are some of the factors of 120.

Making a box from grid paper also gives people the chance to work with area and volume.

Area is simply the size of a surface. In this activity, people measure the length, width, and height of a rectangle in centimeters. When they multiply length and width to get area, their unit of measurement is square centimeters, or cm^2. (The 2 in cm^2 is an *exponent*. For more on exponents, see page 133 or the glossary.)

People also measure the volume of their boxes. *Volume* is the amount of space an object occupies. It's usually defined in math class by a formula (length times width times height). When people multiply length by width by height to get volume, their unit of measurement is cubic centimeters, or cm^3.

The volume of each greeting card box is also a measurement of how much the box can hold (a concept that might be more familiar to your group). If these were wooden boxes with thick walls, the volume *of* the box and the volume *held* by the box might be different measurements. But with these thin-walled paper boxes, they are pretty close to the same.

Because everyone uses the same area for the rectangle they draw on the grid paper, they can see how volume changes when their rectangles' lengths and widths are different. The volume of a box doesn't depend on only the area of the original rectangle. It also depends on the rectangle's dimensions. Rectangles with lengths and widths that are nearly the same make boxes that enclose larger volumes. Long, skinny rectangles (like 5 cm by 24 cm) enclose less volume than boxes whose dimensions are closer together (like 10 cm by 12 cm).

Jacob's Ladder

Jacob's ladder is an intriguing toy made of wooden blocks linked by ribbons. When the top block is flipped over, the second block seems to tumble down the "ladder" of blocks. Your group can make similar toys from paper and ribbon, and then work together to figure out what's really going on when the block tumbles down the ladder.

Preparation and Materials

For each person making a toy, you will need:

- for the paper packets: 18 paper clips, transparent tape, and 6 copies of *Jacob's Ladder Section Sheet* (If you prefer, you can substitute wooden blocks or cardboard rectangles for the paper packets; you will need 6 per person.)

- 15 pieces of flat, smooth, colored ribbon exactly 15 cm long and about $\frac{1}{2}$ cm wide (You can cut ribbons for your group or have each person cut his or her own, which gives valuable practice in measuring in centimeters. We suggest you have people work in pairs to measure and cut the ribbon. Each pair will need 450 cm of ribbon, a pair of scissors, and a centimeter ruler.)

- a copy of *Assembling Jacob's Ladder*

- a copy of *Exploring Jacob's Ladder* and a pencil (optional)

Using This Activity

Your group can make Jacob's ladders on one day and experiment with them on another. If you have an hour or longer, you can do both activities in one afternoon.

More tips on using *Jacob's Ladder* start on page 116.

Planning chart

Cutting ribbons as a group	10 minutes
Assembling Jacob's ladder	45 minutes
Exploring Jacob's ladder	20 minutes
Trying the dollar bill trick	10 minutes

The Math Explorer
Published by Key Curriculum Press / © 2003 Exploratorium

Assembling Jacob's Ladder

PART 1: FOLDING THE PAPER PACKETS

Jacob's ladder is a toy that does some very weird things. The first step in making this toy is folding six paper packets.

What Do I Need?

◇ 6 *Jacob's Ladder Section Sheets*

◇ 18 paper clips

◇ transparent tape

What Do I Do?

Step 1 Fold a *Jacob's Ladder Section Sheet* in half, along the dotted line labeled "first fold," so that the pictures and words are facing out.

Step 2 Fold the sheet in half again, along the dotted line labeled "second fold," so that the black tabs and the paper clip pictures are on the outside.

Step 3 Tape three paper clips on top of the pictures of paper clips.

Step 4 Fold the sheet over the paper clips, along the dotted line labeled "third fold," so that the paper clips cannot be seen. See picture at right.

Step 5 Tape the folded paper around all the edges that aren't folds so that the paper clips are sealed inside.

Step 6 Repeat Steps 1–5 with each *Jacob's Ladder Section Sheet* so that you have a total of six packets that look like the one shown at right.

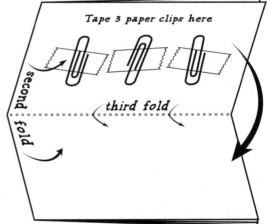

PART 2: ADDING THE RIBBONS

The ribbons are the secret to the strange behavior of the Jacob's ladder toy.

What Do I Need?

◇ 6 paper packets

◇ 15 pieces of flat, smooth ribbon exactly 15 cm long

◇ transparent tape

What Do I Do?

Step 1 There should be three marks on each side of each of your paper packets. Check that all your packets have these marks. They show you where to tape the ribbon pieces.

Step 2 Take one packet, and tape three ribbons on top of the marks. Make sure the ribbons come straight out of the sides, not off at an angle. All three pieces need to be taped to the same side of the packet. Don't tape ribbon to the other side of the packet.

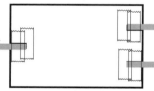

Step 3 Tape three ribbons each to four more packets, as you did in Step 2.

Stop! You should now have five packets with ribbons taped to them. *Don't tape any ribbons to the sixth packet!*

The Math Explorer
Published by Key Curriculum Press / © 2003 Exploratorium

PART 3: PUTTING IT TOGETHER

You're almost done. Now you have to put all the pieces together to make a Jacob's ladder. When you're done, use your toy to mystify your friends.

What Do I Do?

Step 1 Place one packet on the table in front of you, with the ribbon side down. Have two ribbons sticking out to the right and one to the left.

Fold each ribbon back over the card. It will look like the picture shown at right.

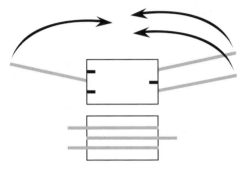

Step 2 Stack the next packet on top, ribbon side down, so that the unused marks on the second packet line up with the short pieces of ribbon sticking out on the sides of the first packet. The long pieces from the second packet will also be sticking out—two to the right and one to the left.

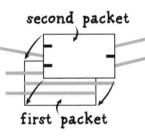

second packet

first packet

Step 3 Fold the three short pieces of ribbon from the first packet back onto the marks on the second packet, and tape them in place on top of the marks. Try to attach the tape close to the edge of the packet.

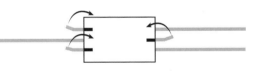

Don't pull the ribbon tight before you tape it! A little looseness makes a Jacob's ladder work better. Now it will look like the picture shown at right.

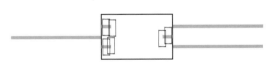

Step 4 Now fold the long pieces of ribbon back over the packet, as you did in Step 1.

Step 5 Repeat Steps 2 through 4 until you have placed and taped all five of the packets that had ribbons attached to them.

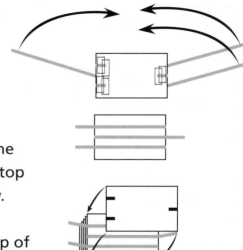

Step 6 The last packet you add to your stack is the one without any ribbons. Place it so that the marks on the top line up with the folded ribbons from the packet below.

Fold back the three short pieces of ribbon that are sticking out on the sides, and tape them in place on top of the last packet.

Step 7 *Now, give your Jacob's ladder a try!* Hold the top packet by its edges, and let the other five packets dangle. The top packet will have one or two ribbons on one side and no ribbons on the other side.

Tip the packet in the direction of the side with ribbons. Keep tipping the packet until its top edge touches the bottom edge of the second packet. When this happens, the second packet will begin to tumble.

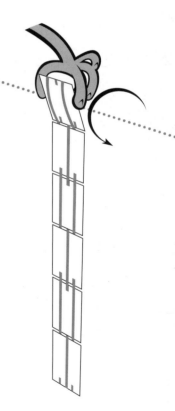

Exploring Jacob's Ladder

Jacob's ladder toys have been puzzling people for hundreds of years. What do you think is going on when you set the toy in motion? Can you think of a way to figure out whether your guess about what's going on is right?

At the Exploratorium, we figured out a few steps to help us see what's going on when we set our Jacob's ladder in motion.

What Do I Need?

◇ Jacob's ladder toy

◇ a pencil

What Do I Do?

Step 1 It's hard to see what happens when you flip your Jacob's ladder over, because all the packets look alike. As the first step in our experimentation, we decided to change that.

Lay your Jacob's ladder flat on a table. Write a number on each packet, starting with 1 at the top and going down to 6. Flip the Jacob's ladder over, and number the other side of each packet. Make sure the packet you label 1 on one side also says 1 on the other side.

Step 2 Pick up the toy by the packet you've labeled 1. Hold the ladder so it hangs, and look at your numbers. Are they right-side up or upside down?

Step 3 What do you think will happen to the packets when you start the cascade? Try it. The numbers will help you see where the packets end up.

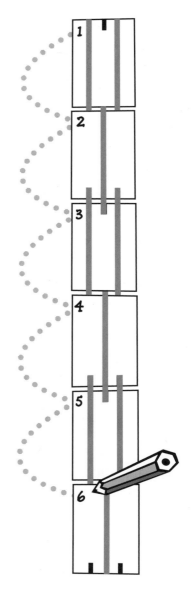

Step 4 Are the packets still in order (1 at the top, followed by 2, and so on)? What is different? Try making the packets tumble again, and watch what happens.

Step 5 Putting numbers on the packets helped us figure out some patterns, but we realized that we also needed to be able to tell one side of a packet from the other. Draw a star on one side of each packet—the side that is facing you right now.

Step 6 Start the cascade again, and watch what happens to the numbers and the stars. Keep playing with your Jacob's ladder until you can predict what will happen each time you begin the cascade.

Step 7 Once we figured out what was happening with the numbers and the stars, we took a close look at the ribbons. Is there a pattern to what happens to them? Make pencil lines around the ribbons before one cascade, and then let the cascade fall to see what happens.

A Trick to Try

Even after you can predict what your Jacob's ladder will do, it's still one weird and puzzling toy. Here's a trick you might want to try.

Fold a dollar bill into a small rectangle, and tuck it under one of the ribbons. Then operate the toy several times. If you watch the Jacob's ladder from one side, the paper money will seem to disappear and then reappear. (That's because, as the packets tumble, the dollar bill switches from one side to the other.)

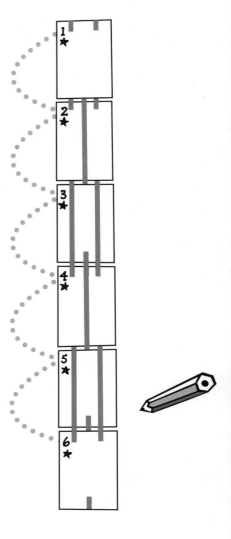

The Math Explorer
Published by Key Curriculum Press / © 2003 Exploratorium

Jacob's Ladder Section Sheet

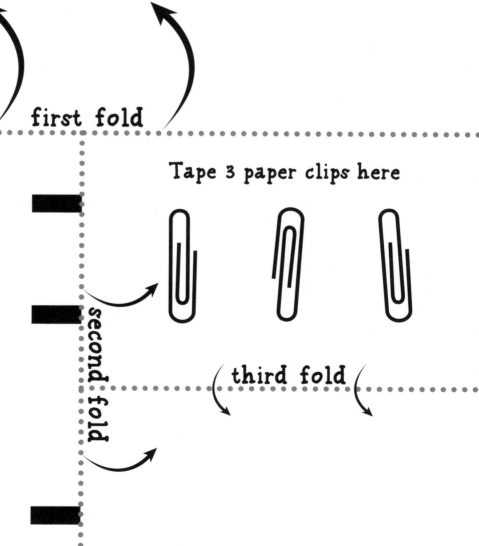

first fold

Tape 3 paper clips here

second fold

third fold

Using *Jacob's Ladder* with Your Group

We strongly suggest you show your group a Jacob's ladder toy—traditionally made of wooden blocks connected by colored ribbons—before having them try this activity. Toy stores and novelty shops sometimes sell them. Or you can make a paper one by following the instructions. This is such an intriguing toy that everyone will want one!

A Short History of This Toy

The Jacob's ladder toy may have originated in China over two thousand years ago. Its popular name came from an African American spiritual, "We Are Climbing Jacob's Ladder." The spiritual is based on the biblical story of Jacob (Abraham's grandson), who dreamed he saw angels walking up and down a ladder between heaven and Earth.

Making a Jacob's Ladder

Instructions for constructing the toy using paper packets instead of wooden blocks appear in *Assembling Jacob's Ladder* beginning on page 109. This project involves three tasks: folding paper packets, taping ribbons to the packets, and assembling the toy. Tell your group that making this toy will take patience, but the payoff will be worth the time.

Folding the Paper Packets

Demonstrate how to fold a paper packet. After the two folds, the side with the pictures of the paper clips should be showing. Real paper clips should be taped on top of the pictures; one piece of tape can hold all three paper clips in place. (The paper clips add a

little weight to the paper, which makes the toy work better.) The last fold covers the paper clips.

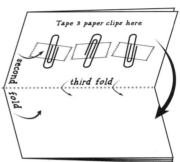

Point out the marks on the outside of the packets. If the marks don't show, the packet has been folded incorrectly. The folded packet should be taped on all edges except the fold.

Each person will need six packets. Ask people who work faster to help others before they move on to the next step.

Using Substitutes for Paper Packets

Other materials can be substituted for the paper packets. Traditionally, Jacob's ladder toys are made of wooden blocks, which make a satisfying "clack clack" noise as they tumble. At the Exploratorium, we have made them from blocks of wood and from cardboard rectangles. Experiment with materials you have on hand, and you may find something you prefer to paper.

Adding the Ribbons

On the folded paper packets, printed marks show where to tape the ribbons. Only five of the six packets will have ribbons taped to them. Suggest that people set one packet aside before they start taping.

It is important that people tape three ribbons to each of five of their packets. The three ribbons must all be taped to the same side of the packet, right on top of the marks. The marks are 1.5 centimeters long, and that is how much ribbon should be taped onto the packet. Again, people who work faster can help the others.

Putting It Together

This is the trickiest part of the process. It's easy for people to get confused and frustrated. Encourage members of your group to take their time and follow the instructions carefully. We suggest that you demonstrate how to do these steps.

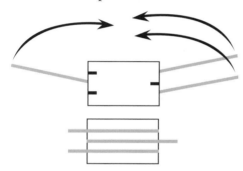

After a demonstration, some people may be able to layer and tape their packets by following the written directions. Continue to demonstrate for those who need to see this again.

Once people understand the process, monitor their progress, checking on each person's work. Again, ask those who finish more quickly to help the others.

After all the packets with ribbons have been used, add the packet with no ribbons to the top of the stack. People need to fold back the long ribbons on the fifth packet before adding the sixth packet. The short pieces of ribbon need to be taped to the top side of the sixth packet.

Giving Your Jacob's Ladder a Try

Follow the instructions to set your Jacob's ladder in motion. If the cascade doesn't start when you rotate the top packet in one direction, try rotating it the other way. If that still doesn't work, shake the whole ladder a little, as the packets can get "stuck." Some people may be unfamiliar with the word *cascade*, which means "to fall in a series of stages."

Exploring *Jacob's Ladder*

Trying to figure out what's happening when a block or packet tumbles down Jacob's ladder can stretch the mind of a person of any age. Experimenting with this toy gives your group a chance to make predictions and look for patterns, essential skills in math and science.

Work in Small Groups

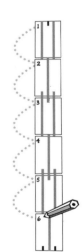

If you are working with a large group, you may want to have them break up into small groups to experiment. The steps in *Exploring Jacob's Ladder* are written so that members of your group can follow them, but each small group will benefit from the guidance and encouragement of a leader as they work through the steps.

Observe and Predict

When scientists set out to understand something, they usually begin with careful observation. The steps suggested in *Exploring Jacob's Ladder* are designed to help people notice what's going on.

After scientists have observed a system, they often make predictions. "What will happen if we do this?" they ask. Making simple predictions—like "I think Packet 1 will end up at the bottom of the ladder"—is an important step in figuring out how this toy works. It doesn't matter whether someone's initial prediction turns out to be right. What matters is making a prediction—and then testing it by experimenting.

Guiding the Experimentation

Ask members of your group what they think is going on when a packet tumbles down Jacob's ladder. Allow time for them to answer before they experiment. Because you are asking them to form hypotheses, there are no "wrong" answers.

Explain that writing numbers on the packets will help people see where the packets end up and how many different positions the packets can be in.

Here are some questions to ask as people experiment:

- Are the packets still in order, with 1 at the top, followed by 2, 3, 4, 5, and 6? (People should see that the packets are in the same order as before.)

- What is different? (People may notice that the numbers have flipped over—if they were right-side up, now they are upside down. If they were upside down, now they are right-side up.)

After experimenting with numbering their packets, people can mark one side of each packet. We suggest stars, but they can use anything they want, such as pictures or words that spell out a message.

Encourage them to keep playing until they can predict what will happen each time

the cascade begins. Each time a packet seems to tumble down the ladder, all the packets flip over. The ribbons on each packet will switch from one on a particular side to two.

But What's *Really* Going On?

Jacob's ladder is great fun to experiment with—but don't be surprised if you still feel a little mystified after playing with it. At the Exploratorium, we've been playing with this toy for a long time, and we still find it puzzling and intriguing. We've baffled mathematicians and scientists with this simple toy.

After experimenting with Jacob's ladder, your group should be able to predict what will happen each time it sends a packet tumbling down the ladder. That's a great start toward understanding a toy that has puzzled people for centuries.

Where's the Math?

Experimenting with this toy hones people's visualization skills, building a spatial awareness important to understanding three-dimensional geometry. Predicting outcomes and experimenting to test those predictions are key problem solving skills. It's important for people to realize that mathematics involves much more than numbers and calculations—it includes strange puzzles like this one!

The Math Explorer
Published by Key Curriculum Press / © 2003 Exploratorium

Paper Engineering

A few cuts and folds can transform an ordinary sheet of paper into an amazing three-dimensional sculpture or pop-up card. Paper engineering encourages experimentation and creativity and, at the same time, helps develop spatial awareness.

Preparation and Materials

For each person, you will need:

- a few $8\frac{1}{2}$-by-11-inch sheets of colored paper

- scissors

- a centimeter ruler (If you don't have rulers, see page 202.)

- a pencil

- a few $8\frac{1}{2}$-by-11-inch sheets of card stock (white or a different color from the paper)

- a glue stick

- copies of *Paper Engineering with Fractals*, *More Paper Engineering*, and *Still More Paper Engineering*

If you have time, make a fractal sculpture in advance to show people what they'll be making. Simply follow the directions on *Paper Engineering with Fractals*. Completing the sculpture will take 10–15 minutes.

Using This Activity

Have your group make pop-up cards for Mother's Day, Valentine's Day, Christmas, or any other holiday that involves giving cards. If you don't have enough colored paper for *More Paper Engineering*, you can substitute colorful ads and pictures cut out from magazines.

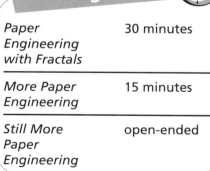

Planning chart

Paper Engineering with Fractals	30 minutes
More Paper Engineering	15 minutes
Still More Paper Engineering	open-ended

Paper Engineering with Fractals

Learn the basics of paper engineering by making this three-dimensional sculpture from a flat sheet of paper. Then put your paper-engineering skills to work to create amazing pop-up cards and sculptures.

What Do I Need?

◇ an $8\frac{1}{2}$-by-11-inch sheet of colored paper

◇ scissors

◇ a centimeter ruler

◇ a pencil

◇ an $8\frac{1}{2}$-by-11-inch sheet of card stock (white or a different color from the paper)

◇ a glue stick

What Do I Do?

Look at the picture for each step to make sure you have it right before you move on.

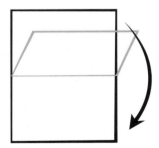

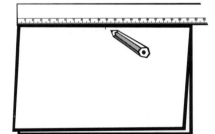

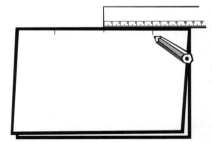

Step 1 Fold your sheet of colored paper in half.

Step 2 Place your ruler along the fold you just made, and find the center of the paper along the fold. Make a small pencil mark there.

Step 3 Starting at the center mark, measure 6 centimeters (cm) along the fold in each direction. Make a mark at each of these measurements.

The Math Explorer
Published by Key Curriculum Press / © 2003 Exploratorium

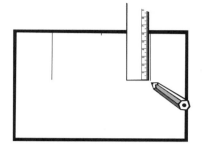

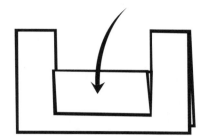

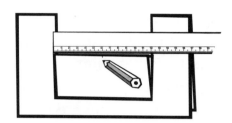

Step 4 At each of these marks, use your ruler to draw a straight line that's 6 cm long, perpendicular to the fold. (Perpendicular means the line meets the fold at a 90-degree angle.)

Step 5 Cut along these two lines to make a flap. Fold this flap toward you and down.

Crease the paper along the fold. Now fold the flap away from you and crease it again.

Step 6 Use your ruler to find the center mark along the fold you just made. Make a mark there. Be sure to measure and mark on the top fold of the flap, as shown.

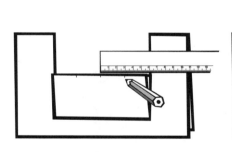

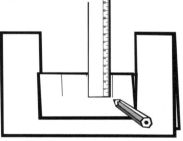

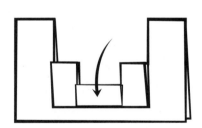

Step 7 Measure 3 cm from the new center mark, along the fold. Do this on each side of the new center mark.

Step 8 At each of these marks, draw a straight line that's 3 cm long, perpendicular to the fold.

Step 9 Cut along the 3-cm lines to make a flap. Fold this flap toward you and down.

Crease the paper along the fold. Now fold the flap away from you and crease it again.

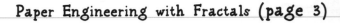
Do you get the feeling you've done this before? You are repeating the same set of actions. A computer programmer might call this an *iterative procedure* that is also *recursive*. That's just a fancy way of saying that you are following the same instructions again and again, each time starting where the last step ended.

Step 10 Time for one more *iteration*. (That means we're going to do the same thing one more time.) Find the center mark on the newly folded section.

Step 11 The first time you measured 6 cm from the center. The second time you measured half that distance, just 3 cm. This time you will measure half of 3 cm. Measure 1.5 cm from the center mark, along the newest fold, in each direction. Make a mark at each spot.

Step 12 At each of these marks, draw a straight line that's 1.5 cm long, perpendicular to the fold.

Step 13 Cut along each of these lines to make a flap. Fold this flap toward you and down, as shown.

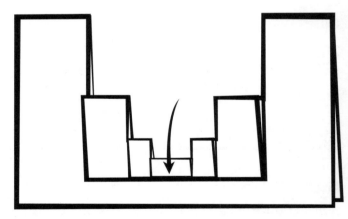

Crease the paper along the fold. Now fold the flap away from you and crease it again. It's probably getting hard to crease the paper, but do your best to make sharp creases. Later on, you'll be glad you did.

Step 14 Carefully unfold all the folds, starting with the most recent. When you have all the folds undone, you'll have a piece of paper with a lot of cuts and folds in the middle. It doesn't look too spectacular . . . yet.

The Math Explorer
Published by Key Curriculum Press / © 2003 Exploratorium

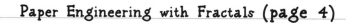
Step 15 This is the tricky part. You've got lots of folds and lots of cuts. These folds and cuts make a set of squares—with cuts on the sides of each square and with folds at the top, the middle, and the bottom of each square.

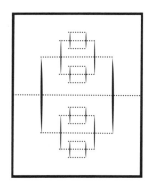

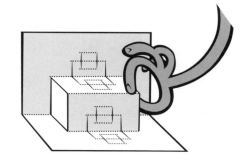

Start with the largest square. Pull up on the middle fold of this square as you fold the whole sheet of paper in half. The flap will pop up to make a shelf.

Step 16 One by one, push each of the smaller squares from behind as you fold the sheet. Make each one of them pop out to make a shelf.

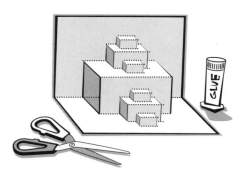

Voila! (That's French for "There it is!") You've got a three-dimensional fractal sculpture. Fold the sheet of card stock in half and put it behind your sculpture. Glue the sculpture to the card stock.

What's a Fractal, Anyway?

Fractals are patterns that repeat at different scales. Take the fronds of a fern, for example. Each frond sprouts miniature copies of itself. These copies, in turn, sprout even smaller versions of the same shape. This repetition at smaller and smaller sizes is a characteristic of fractals.

More Paper Engineering

Now that you know the basics, you can experiment and make some other cool pop-up sculptures.

What Do I Need?

◊ a few $8\frac{1}{2}$-by-11-inch sheets of colored paper

◊ scissors

◊ a centimeter ruler

◊ a pencil

◊ a few $8\frac{1}{2}$-by-11-inch sheets of card stock (white or a different color from the paper)

◊ a glue stick

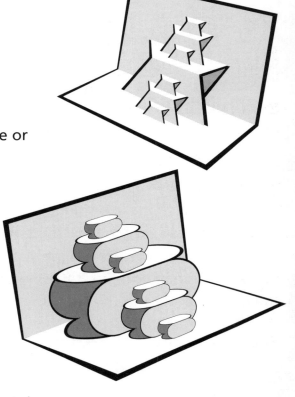

What Do I Do?

Step 1 Here are two examples of fractal sculpture that folks at the Exploratorium made.

Can you figure out how to make a fractal sculpture like one of these? (Here's a hint: the cuts you make don't always have to be straight or perpendicular to the fold.)

Step 2 Think of how you might change your fractal sculpture to make it more interesting. You could try making cuts that aren't perpendicular to the fold. When you cut the sides of a flap, you could make one cut longer than the other, so that the fold of the flap is at an angle.

Step 3 Experiment. Try doing something a little different, and see how it affects your completed sculpture.

Step 4 When you make something you really like, fold a sheet of card stock in half and put it behind your sculpture. Glue the sculpture to the card stock.

The Math Explorer
Published by Key Curriculum Press / © 2003 Exploratorium

Still More Paper Engineering

Here are some ideas that you can use alone or with your fractal sculpture to make more unusual pop-up cards.

Making a Pop-Up Beak

Step 1 Fold a sheet of paper in half.

Step 2 Make a cut perpendicular to the fold. The length of this cut is up to you.

Step 3 Fold a triangle on either side of the cut, as shown here.

Crease the fold on each triangle. Now fold each triangle to the other side of the card, and crease this fold, too.

Step 4 Unfold both triangles. Open the sheet of paper as if you were opening a greeting card. Put your fingers on the back of the triangles and push them toward the front, letting the card fold shut a little as you push. When you open and shut the card, the beak will open and close.

Step 5 If you want, you can draw eyes above the beak or make a word balloon coming out of the beak. Glue a sheet of card stock behind your creation to make a greeting card.

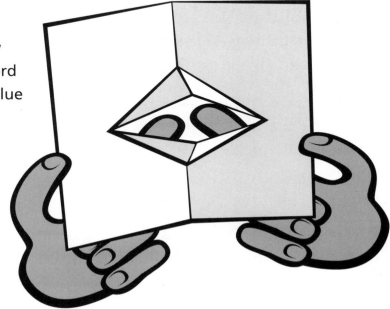

Making Pop-Up Lips

Step 1 Follow Steps 1 and 2 of "Making a Pop-Up Beak."

Step 2 Figure out how wide you want the lips to be. Measure exactly that far on either side of the cut you made, and make a mark.

Step 3 Draw a line that's perpendicular to the fold at each of these marks. This line should be about three-quarters the length of your first cut. You don't have to be exact.

Step 4 Cut along each of these lines.

Step 5 Fold the lips out.

Step 6 Unfold the lips. Open the sheet of paper as if you were opening a greeting card. Put your fingers on the back of the lips and push them toward the front, letting the card fold shut a little as you push. When you open and close the card, the lips will open and close, too.

Step 7 If you want, you can draw eyes above the lips or draw a word balloon. Glue a sheet of card stock behind your creation to make a greeting card.

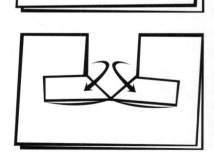

Have fun!

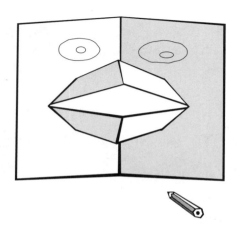

The Math Explorer
Published by Key Curriculum Press / © 2003 Exploratorium

Using *Paper Engineering* with Your Group

Creating pop-up sculptures and cards is called *paper engineering*, and it's a respected profession in the publishing business.

In *Paper Engineering with Fractals*, each member of your group will make a pop-up sculpture to take home. Making this sculpture requires careful measurement, something that most middle schoolers need to practice. The sculpture relates to fractal geometry, an interesting area of mathematics.

More Paper Engineering and *Still More Paper Engineering* give members of the group a chance to experiment with other sculptures and pop-up cards.

Making Pop-up Sculptures

If members of your group follow written directions well, you can give them copies of *Paper Engineering with Fractals* (page 120) and let them work alone or in pairs. If you want to keep the group together or if members of your group don't follow written directions well, you may want to demonstrate each step of the process.

When your group moves on to *More Paper Engineering*, you may need to demonstrate some of the possible ways to change the fractal sculpture. One of the simplest is to make cuts at an angle other than perpendicular to the fold, as was done with the design shown above right.

This is a chance for group members to experiment and make something they like. Encourage them to experiment. If an experiment doesn't work, that's no big deal. It's only paper. Simply try something else.

Sometimes, people find it hard to figure out what they'll get when they are done cutting and folding a pop-up sculpture. But if they keep experimenting, chances are they'll get better at predicting what will result from a particular set of cuts and folds.

Making Pop-Up Cards

Encourage group members to make use of what they learned in doing *Paper Engineering with Fractals* and *More Paper Engineering* as they make pop-up cards. Can someone make a fractal sculpture that has many pop-up beaks, each with its own word balloon? Give it a try!

Stephanie Su of San Francisco's Community Youth Center said that this activity really brought out her group's creativity as members experimented with pop-up shapes.

For More on Fractals

If group members are interested in learning more about fractals and exploring computer-generated fractal patterns, have them check out some interesting links at

www.keymath.com/ME

Where's the Math?

Before 1970, mathematicians found it almost impossible to describe natural shapes—like the billows of a cloud or the crags of a mountain. Regular geometry—which deals with straight lines and smooth curves—couldn't cope with these irregular shapes. Nineteenth-century mathematicians called these troublesome forms "mathematical monsters" and avoided dealing with them.

In the 1970s, a mathematician named Benoit B. Mandelbrot developed fractal geometry and gave mathematicians a weapon for dispatching these mathematical monsters. The billows of a cloud, the branches of a tree, the irregular crags of a mountain—all these are fractal patterns that can be described by fractal geometry.

Fractal geometry has led to the creation of fractal forgeries—computer-generated images that look like forms in the natural world. A computer builds a fractal image in the same way that you made your fractal design sculpture—by following the same instructions over and over again.

When a computer makes a fractal pattern, it plugs numbers into a mathematical formula. It makes a calculation, plots a point, and plugs the results of the first calculation back into the formula. It then repeats the calculation. This kind of repetition is called *recursion* or *iteration*.

In making your sculpture, you followed the same instructions three times. To make a fractal image of a cloud or a tree or a landscape, a computer follows the same instructions hundreds of thousands of times.

The Math Explorer
Published by Key Curriculum Press / © 2003 Exploratorium

Incredible Shrinking Shapes

Your group can shrink plastic recycled from deli containers to make medallions and earrings. In the process, people learn about area, volume, and ratios, important concepts in middle school math.

Preparation and Materials

For this activity, you will need:

- a toaster oven or standard oven heated to 350°F (170°C) in a well-ventilated area, a cookie sheet or aluminum foil, a spatula, and a pot holder

- polystyrene (recyclable number 6) plastic (1 piece per person) (*Do not use other types of plastic;* they may emit toxic fumes when heated.)

- fine-tipped permanent markers, preferably in assorted colors (at least 1 per person)

- a hole punch, scissors (1 pair per person), metric rulers (1 per person), and calculators (enough to share)

- copies of *Incredible Shrinking Shapes Data Sheet* and *Grid Paper* (see page 204)

If you want your group to determine how the thickness of the plastic changes when it shrinks (see page 136), each person will also need a clothespin or a binder clip.

Using This Activity

Obtain the plastic by collecting clear-plastic salad bar and deli containers and clear yogurt lids. To make sure you've got the right type, look for this symbol:

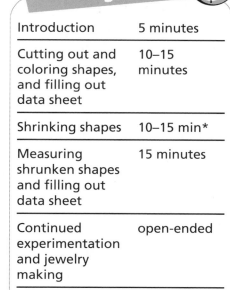

Start collecting plastic a few weeks in advance. If you need more than you can collect, you can buy clear plastic containers at restaurant supply stores.

More tips on how to use *Incredible Shrinking Shapes* start on page 133.

Planning chart

Introduction	5 minutes
Cutting out and coloring shapes, and filling out data sheet	10–15 minutes
Shrinking shapes	10–15 min*
Measuring shrunken shapes and filling out data sheet	15 minutes
Continued experimentation and jewelry making	open-ended

* The time it takes to shrink all the shapes will depend on the size of your group and the size of the oven. You can fit 3 or 4 pieces in a toaster oven at one time. They need to be in for about 1 minute.

Incredible Shrinking Shapes Data Sheet

Make cool jewelry from plastic deli containers!

What Do I Do?

Step 1 Trim away any parts of your plastic that aren't flat. Lay your plastic on the grid paper. The grid is marked off in square centimeters (cm²).

Step 2 Using a ruler and marker, trace the lines of the grid to draw two identical squares or rectangles. Make them at least 4 cm by 4 cm but smaller than 8 cm by 8 cm. Cut out your shapes.

Step 3 Measure the length and width of your shape:

    Length = _____ **cm**   **Width =** _____ **cm**

Step 4 Calculate the area of the shape:

$$\underset{\text{(length)}}{\underline{\hspace{2cm}}} \text{ cm} \times \underset{\text{(width)}}{\underline{\hspace{2cm}}} \text{ cm} = \underset{\text{(area)}}{\underline{\hspace{2cm}}} \text{ cm}^2$$

Step 5 Decorate your shapes using permanent markers. If you want to make pendants, earrings, or zipper pulls, make holes in the shapes using a hole punch.

Step 6 Take one of your shapes to the oven, where the group leader will shrink it.

Step 7 Compare your shrunken shape to the unshrunken shape. Measure the length and width of your shrunken shape. Then calculate its area:

    Length = _____ **cm**   **Width =** _____ **cm**

$$\underset{\text{(length)}}{\underline{\hspace{2cm}}} \text{ cm} \times \underset{\text{(width)}}{\underline{\hspace{2cm}}} \text{ cm} = \underset{\text{(area)}}{\underline{\hspace{2cm}}} \text{ cm}^2$$

The Math Explorer
Published by Key Curriculum Press / © 2003 Exploratorium

Step 8 Now compare the length of your shape before shrinking with the length after shrinking by setting up a *ratio*. Fill in the circles in this equation:

$$\frac{\textbf{Length before shrinking}}{\textbf{Length after shrinking}} = \frac{\bigcirc \ \ \textbf{cm}}{\bigcirc \ \ \textbf{cm}} = \bigcirc$$

A ratio is a way of comparing two numbers.

$$\frac{?}{?}$$

Divide the *length before shrinking* by the *length after shrinking*. Round your answer to the nearest tenth. Write the answer in the last circle in the equation. This is your *length ratio*.

Your length ratio is probably between 2 and 3. If your length ratio is 2, it means the original shape was 2 times the length of the shrunken shape.

Step 9 Now compare the width of your shape before shrinking with the width after shrinking. Fill in the circles in this equation:

$$\frac{\textbf{Width before shrinking}}{\textbf{Width after shrinking}} = \frac{\bigcirc \ \ \textbf{cm}}{\bigcirc \ \ \textbf{cm}} = \bigcirc$$

Divide the *width before shrinking* by the *width after shrinking*. Round your answer to the nearest tenth to get the *width ratio*.

Published by Key Curriculum Press / © 2003 Exploratorium

Step 10 You know what happened to the length and width of your shape. Now you will find out what happened to the area.

Set up one more ratio, comparing the *area before shrinking* with the *area after shrinking:*

$$\frac{\textbf{Area before shrinking}}{\textbf{Area after shrinking}} = \frac{\bigcirc\ \text{cm}^2}{\bigcirc\ \text{cm}^2} = \bigcirc$$

Divide the *area before shrinking* by the *area after shrinking.* Round the answer to the nearest tenth. Write your result in the last circle in the equation. This is your *area ratio.*

If your area ratio is 7, it means your original shape had an area 7 times the area of your shrunken shape.

Compare your shrunken shape with your unshrunken shape. How many shrunken shapes would it take to completely cover the unshrunken shape? That number should be the same as your area ratio.

Step 11 Which is greatest—your area ratio, your length ratio, or your width ratio? Why might this be? Talk with your group about this.

Step 12 If there's time, shrink your second shape—and experiment with other shapes.

Making *Incredible Shrinking Shapes*

This activity gives people hands-on experience with ratios and scaling, important concepts in middle school math. At the same time, members of your group can make their own jewelry

Before You Begin

A Note of Caution

Start by preheating the oven to 350°F (170°C). Do not turn the oven up any higher. Cooking this type of plastic is considered safe at 350°F, but toxic fumes may be released if the plastic is exposed to temperatures of 400°F or higher. For safety's sake, do this activity in a well-ventilated area.

A Chance to Play

Jeanne D'Arcy, a teacher who runs an after-school program, says that her group loved this activity. Rather than starting with the math, she gave her group a chance to experiment with the materials and to make some shrinking shapes before she introduced the concept of ratios. Jeanne also suggests putting the materials needed for this activity in different areas so that people have to walk around the room to get what they need. After a long day at school, many students don't want to sit still.

Introduction of Square Centimeters

If you haven't done any activities involving centimeters (cm), you might want to mention that the lines on the grid paper are spaced 1 centimeter apart.

Where's the Math?

When centimeters are multiplied by centimeters, the result is square centimeters, or cm^2, which are used to measure area. The 2 is an *exponent*. An exponent is a small number written above and to the right of another number or term that is called the *base*. The exponent indicates how many times the base is used as a factor. The unit cm^2 means this:

$$cm \times cm$$

The unit cm^3 means this:

$$cm \times cm \times cm$$

Cubic centimeters, or cm^3, are used to measure volume. Exponents can be used with numbers as well. For example:

$$5^2 = 5 \times 5$$

$$5^3 = 5 \times 5 \times 5$$

$$5^4 = 5 \times 5 \times 5 \times 5$$

Ask your group what units they get when they multiply centimeters by centimeters. (They get square centimeters, or cm^2.)

Cutting Out Shapes

When your group is working with the *Incredible Shrinking Shapes Data Sheet*, we suggest having everyone make a square or a rectangle. It's tough to calculate the areas of irregular shapes.

Shrinking the Shapes

To shrink the shapes, place them on a clean cookie sheet or a clean, flat sheet of aluminum foil doubled over. Make sure no shape is touching any other shape. (If they touch, they may melt together.) Shrink only one shape per person.

Put the shapes in the oven, preheated to 350°F. *Watch the shapes carefully*. In less than a minute, they will curl and shrink, and then uncurl.

When the shapes have uncurled, take them out of the oven using the pot holder. Remove them from the cookie sheet or aluminum foil with the spatula. If they are still a bit curled, press them flat between a countertop and a pot holder as soon as you remove them from the oven. Let them cool for a minute before returning them to their owners. Let the cookie sheet or foil cool for a few minutes between batches.

Using the Data Sheet

How Advanced Is Your Group?

Some middle school students are comfortable with the concepts of area and ratios; others have not yet been introduced to these concepts. If your group needs practice measuring and needs to learn about area, we suggest that you distribute only the first page of the *Incredible Shrinking Shapes Data Sheet*.

If your group already understands what area is and how to calculate it, consider giving out all three pages and introducing the concept of ratios.

About Area

Area is a measure of the size of a surface, using square units such as square inches or square centimeters. The *Incredible Shrinking Shapes Data Sheet* explains how to calculate the area of a rectangle: multiply its length by its width.

Area takes two dimensions—length and width—into account. That can make comparing the size of the shrunken shape with the size of the original shape a little tricky.

Ask your group to compare the length of the original shape to the length of the shrunken shape. The original shape probably has a length that's 2 to 3 times that of the shrunken shape.

Ask your group to compare the width of the original shape with the width of the shrunken shape. The original shape probably has a width that's also 2 to 3 times that of the shrunken shape.

Here comes the tricky part. Ask your group to compare the area of the original shape with the area of the shrunken shape. People might think that a square that is twice as tall and twice as wide as another square will have twice the area—but that's not so! Area takes both length and width into account—and so a square that is twice as wide and twice as tall as another square has 4 times the area!

This is an important concept for middle schoolers. You can show your group how this works by using an ordinary rectangular piece of paper as shown in the following steps.

Step 1 Fold the paper in half in one direction, like this:

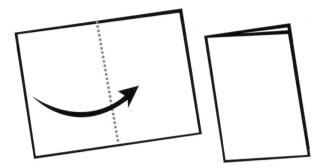

Step 2 Ask the group how many copies of the folded sheet will fit onto the original sheet without overlapping. Show the group that you can fit two folded rectangles onto the original sheet. The original sheet has twice the area of the folded sheet:

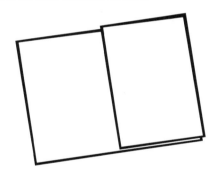

Step 3 Now fold the folded rectangle once more, like this:

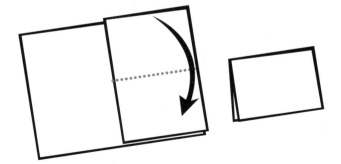

Step 4 Ask the group how many copies of your new rectangle will fit onto the original sheet without overlapping. Show the group that you can fit four folded rectangles onto the original sheet. The original sheet thus has 4 times the area of the twice-folded sheet:

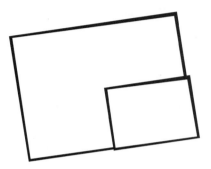

Each time you folded the rectangle, you divided a side by 2. You made the rectangle half as long. Then you made it half as wide. The area of the final rectangle is one fourth that of the original rectangle.

About Ratios

If your group is ready, you can introduce the concept of ratios using the second and third pages of the *Incredible Shrinking Shapes Data Sheet*. A *ratio* is a way of comparing two numbers. In *Incredible Shrinking Shapes*, ratios let people give a quantitative answer to this question: How much smaller is the shrunken shape?

Members of your group can compare the length of the original shape with the length of the shrunken shape. If they use calculators to find the ratio, you may have to talk to them about rounding their answers. (For a discussion of rounding, see page 183.)

The resulting ratio reveals the length of the original shape compared with that of the shrunken shape. The length of the shrunken shape, multiplied by the length ratio, gives the length of the original shape.

This ratio has no units of measurement. When you divide centimeters by centimeters, the centimeters cancel out. The length ratio will probably be between 2 and 3.

People can also compare the width of the original shape with the width of the shrunken shape. This type of plastic often shrinks a little more in one direction than the other. But, generally, the width ratio will be similar to the length ratio—between 2 and 3.

Finally, people can compare the area of the original shape with the area of the shrunken shape. Ask group members to look at their ratios. Can they figure out some mathematical relationship that lets them use the length ratio and the width ratio to get the area ratio?

If they're stumped, ask them how they calculated the area of their rectangle. Area is length multiplied by width. And the area ratio is the length ratio multiplied by the width ratio.

Where's the Science?

Where Did the Plastic Go?

If you have time, ask group members where they think the plastic went when the shapes shrank.

Suggest that they compare the shrunken shapes with the original plastic. Is the new shape thicker? How much thicker? Members of your group can answer this question by stacking pieces of unheated plastic, clamping them together with a clothespin or a binder clip, and comparing the thickness of this stack of plastic with that of the shrunken pieces.

If someone's area ratio is 7, they should find that seven pieces of stacked, unshrunken plastic have about the same thickness as their shrunken piece.

If your group has access to an accurate scale, people can weigh the plastic before and after heating. They will find that its weight doesn't change. One of the laws of nature is that matter can't be created or destroyed. In this activity, matter changes shape, but it doesn't go away. Matter never goes away.

Why Does the Plastic Shrink?

Plastic is made up of long chains of molecules. Up close, these molecules look like a lot of spaghetti. But unlike spaghetti, each chain of molecules is attached to other chains at certain points.

When this plastic was made into a salad tray or a deli container, it was heated and stretched thin over a mold. In that process, the long chains were pulled and stretched, but they were still attached to each other. Heating the plastic in the oven lets the chains "unstretch" and pull back together again. This causes the plastic to shrink.

The Math Explorer
Published by Key Curriculum Press / © 2003 Exploratorium

Looking Up

Exciting a group of young people about measurements, angles, and math can be tough. But it's easy to interest them in launching rockets, flying kites, and making paper airplanes.

In this section, you'll find instructions for making weird-looking kites, constructing rockets that are launched by stomping on a soda bottle, and setting up a paper airplane contest. Math is employed to measure how high the rockets and kites fly and to make the paper airplane contest fair.

Activity 17 **Height Sight**

People build inclinometers, tools that let them measure the height of a distant object. This activity leads into Stomp Rocket! *and* Tetrahedral Kites.

Activity 18 **Stomp Rocket!**

Members of the group make and launch rockets. Using the tools and techniques from Height Sight, *they calculate the height of each rocket's flight.*

Activity 19 **Tetrahedral Kites**

With tissue paper and drinking straws, group members make unique-looking kites. Flying the kites and measuring how high they fly make use of the tools and techniques from Height Sight.

Activity 20 **Flying Things**

People use math to eliminate the natural advantage that tall people have in flying paper airplanes—and to make a paper airplane contest fair.

Height Sight

Inclinometers are devices that let people measure the height of a very tall object. Making and learning how to use them prepares your group for some exciting activities—such as making rockets (page 149) and kites (page 162), and measuring how high they fly. If you live north of the equator, group members can also use inclinometers to figure out where on Earth they are.

Preparation and Materials

For each person, you will need:

- a washer, a 3-by-5-inch card, a pencil, and a sheet of $8\frac{1}{2}$-by-11-inch paper (Reused is fine.)

- 50 cm of string and a piece of string that's about as long as the person is tall

- copies of *Making Your Inclinometer*, *Using Your Inclinometer*, *Height Calculator*, and *Height Calculator Grid*

For the group, you will also need:

- scissors (at least 1 pair for every 3 people), transparent tape, a hole punch, and rulers (at least 1 for every 3 people)

- copies of *Protractors for Inclinometers*

- a way to measure distance (see below)

Using This Activity

If you have time, make an inclinometer before introducing this activity to your group. It will take only a few minutes.

People will need to measure the distance from where they are standing to the object they are measuring. It's easy to do this if your group has done *Stride Ruler* (page 188). Otherwise, members can use measuring tape or measure the distance with string and then measure the string with a meterstick.

More tips on how to use *Height Sight* start on page 145.

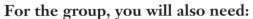

Planning chart

Making inclinometers	15 minutes
Using inclinometers	15 minutes
Calculating heights of objects	15 minutes

Making Your Inclinometer

Here's how to make a tool you can use to measure how tall something is—or how high a rocket or kite flies!

What Do I Need?

◇ a paper protractor, a 3-by-5-inch index card, a sheet of paper, 50 cm of string, and a washer

◇ scissors and a hole punch

What Do I Do?

Step 1 Cut out a protractor from page 144. Cut very carefully along the straight line at the top edge of the protractor.

Step 2 Tape the protractor to the index card so that the straight edge exactly matches up with a long side of the card.

Step 3 Punch a hole through the circle on the protractor.

Step 4 Push one end of the string through the hole and then through the washer. Tie the ends together to make a loop on which the washer can slide freely.

Step 5 Roll the paper into a cylinder $8\frac{1}{2}$ inches long and about 1 inch across. Tape the seam so the paper stays rolled, and then tape the cylinder to the card along the straight edge of the protractor. One end of the cylinder should line up with the side of the card.

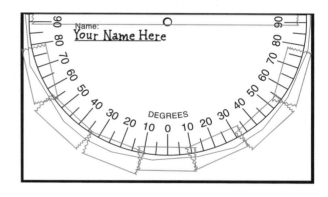

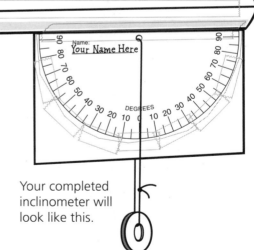

Your completed inclinometer will look like this.

Using Your Inclinometer

Your inclinometer measures how much you have to tilt your head to see the top of something tall. That information lets you figure out how tall that object is.

What Do I Do?

Step 1 Look through the end of the tube that sticks out from the card. You can either look through the tube or sight along the top of the tube. It's your choice.

Step 2 Look at something that's at eye level. Ask a friend to read the angle where the string crosses the protractor. If your inclinometer is level, the string should cross the protractor at about zero degrees.

Step 3 Look through the tube at the top of something tall. If you are indoors, look at something that's near the ceiling. Ask your friend to read the angle where the string crosses the protractor.

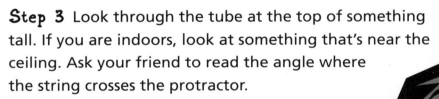

When you look at something above your head, the inclinometer tilts. The string crossing the protractor marks the angle of the tilt.

Take a few steps toward the tall thing, and ask your friend to check the angle on your inclinometer again. What happens to the angle?

Step 4 Rather than having your friend read the angle for you, look through the tube and pinch the string against the card to hold it in place. Then take the tube away from your eye and read the angle on the protractor.

Have your friend watch you do this and tell you if you move the string and change the angle. Try this a few times until you can do it without changing the angle.

Height Calculator

You can use the angle measured by your inclinometer to figure out the height of something tall.

What Do I Need?

◊ an inclinometer

◊ something tall to measure

◊ measuring tape or a long piece of string

◊ a partner

◊ a piece of string about as long as you are tall

◊ a ruler or meterstick

◊ a copy of *Height Calculator Grid*

◊ a pencil

What Do I Do?

Step 1 Choose something tall to measure.

The object I am measuring is _____ .

Step 2 Look at the top of the object through your inclinometer. Determine what angle your inclinometer measures.

The angle on my inclinometer is _____ degrees.

Step 3 Measure the distance (in centimeters) from where you are standing to the base of the object.

The distance to the object is _____ cm.

Step 4 Have a friend measure how far it is from the floor to your eye level. Your friend can measure the distance with string, and then you can measure the string.

The distance from the floor to my eye level is _____ cm.

Step 5 Take a look at the *Height Calculator Grid* on page 143. On the protractor at the lower left corner of the grid, mark the angle you measured in Step 2. Use your ruler to draw a line that runs through the mark on the protractor and through the lower left corner of the grid. Extend this line all the way across the grid.

Step 6 Imagine that your eye is right by the protractor in the lower left corner of the grid. The line at the bottom of the grid is a line parallel to the ground, right at the level of your eyes. Each division on the grid represents 100 cm in the real world. Using this information and the distance you measured in Step 3, mark the object's position along the bottom line of the grid.

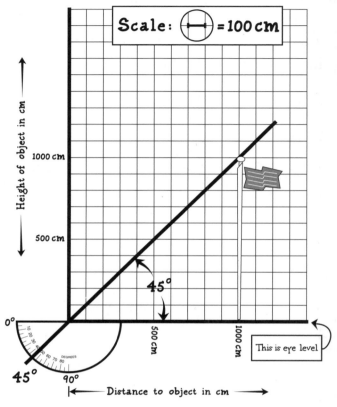

Step 7 Starting at the mark you just made, draw a straight line that's perpendicular to the bottom line of the grid. Draw this line all the way up to the top of the grid.

You've now drawn a triangle. Its height tells you the distance from your eye level line to the top of the object.

The height of the triangle is _____ cm.

Step 8 To get the height of the object, add the distance from the ground to your eye level (from Step 4) to the number you got above.

The height of the object is _____ cm.

Height Calculator Grid

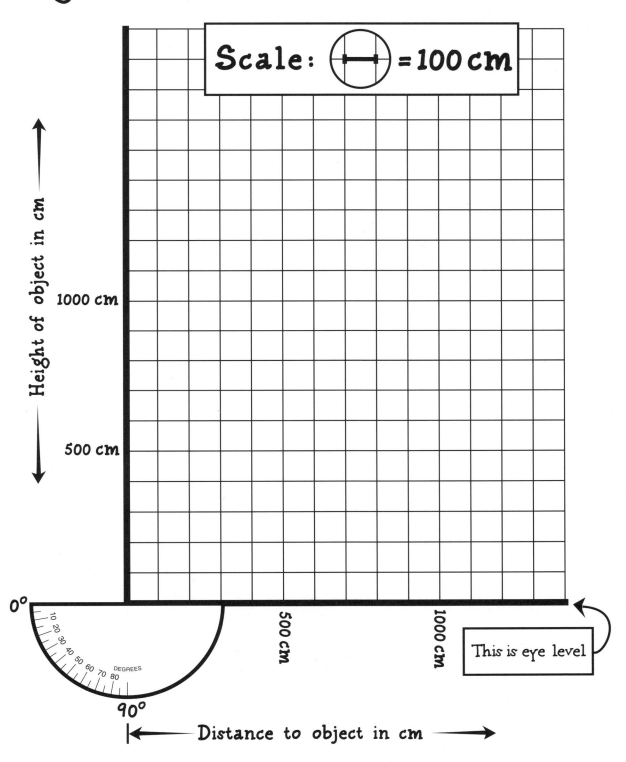

Scale: ⊢—⊣ = 100 cm

Height of object in cm

1000 cm

500 cm

0°

10 20 30 40 50 60 70 80

DEGREES

90°

500 cm

1000 cm

This is eye level

← Distance to object in cm →

Protractors for Inclinometers

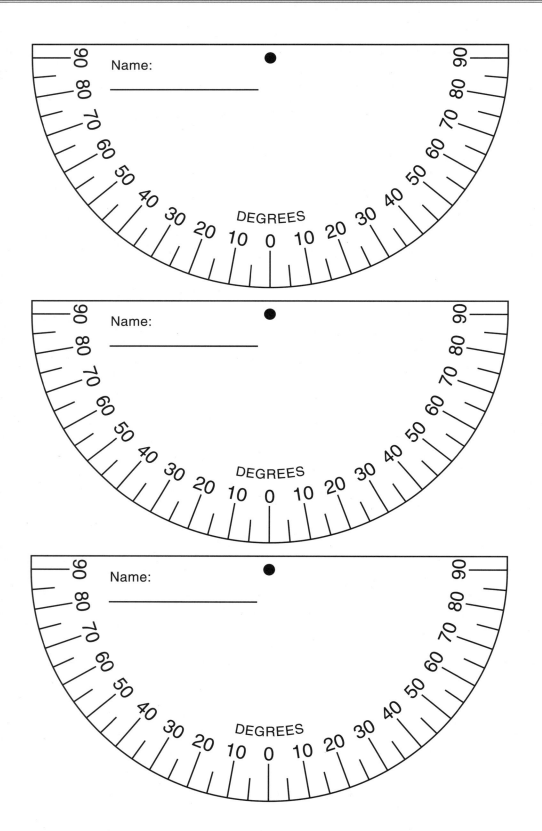

Making and Using Inclinometers

Using an inclinometer, a person can calculate the height of something that he or she can't easily measure with a ruler. This will come in handy when your group does the *Stomp Rocket!* activity (page 149). It can be tough to get a group of young people excited about measurements, angles, and math. But it's easy to get them interested in launching rockets (page 149) and flying kites (page 162). For your group members, the incentive to make and practice using an inclinometer is that they will later use it to measure the height of a rocket's flight.

Making an Inclinometer

Before you have group members start making inclinometers, we suggest you make and show them an inclinometer, and talk a little about it. Ask them to break down the word *inclinometer* to try to figure out what it means. Then tell them that the inclinometer is a tool that measures the height of something too tall to measure directly.

Here are a few questions you might want to ask your group:

- An inclinometer has a protractor on it. What does a protractor measure? (angles)

- If you are planning to build rockets, you might ask group members how they could measure how high a rocket flies. For example, you could compare its height to nearby buildings that you know the height of. But what if there isn't a building nearby? Or what if your rocket goes higher than a building? The inclinometer uses math to measure the rocket's flight.

To make an inclinometer, follow the directions in *Making Your Inclinometer* (page 139). You can copy these instructions for your group or explain the steps one by one while your group follows along.

If one person finishes before the others, you might ask him or her to help people who are working more slowly.

Using an Inclinometer

You can have members of your group follow the steps on copies of *Using Your Inclinometer* (page 140), or you can demonstrate each step while they follow along.

When you demonstrate how to use an inclinometer, look through the end of the tube that sticks out from the card. Otherwise, the weight will swing back and tap you in the face, amusing your group. You can either look through the tube or place the tube on your cheek and sight along the top of it.

Distance Matters

When people use their inclinometers, they may notice that the angles they read on the inclinometer depend partly on how far they are from the object they are measuring. To make sure they understand this, you might want to have two people sight on the same tall object—with one person standing several feet closer to it than the other. The closer person will read a larger angle. Calculating the height of an object requires knowing both the angle and the distance to the object.

Try Both Ways to Read the Angle

We suggest that people work with partners, with one person sighting on the object being measured while the other reads the angle. People should also practice reading the angle without a partner. To do this, they pinch the string against the card to hold it in place. This takes a little practice. Be sure everyone tries this method.

What If People Get Different Answers?

Have the whole group stand the same distance from a tall object, sight on the top of the object, and read the angle using one of the two methods described above. Have everyone compare their angles.

Even if people are the same distance from the object, there will be slight variations among the readings—but they should be within 5–10 degrees of each other. It's normal to have some variation—that's what scientists call *experimental error*. (In the *Stomp Rocket!* activity, we suggest that you have three people measure the height of each rocket's flight. By averaging the three readings, you'll get a more accurate result.)

Calculating an Object's Height

On the *Height Calculator Grid*, people draw a triangle that has the exact shape as the same triangle in the real world.

Suppose one of the people in your group measured a flagpole and drew a diagram like

the one at the right. At the lower left corner of the triangle is the measurer's eye. At the top corner of the triangle is the top of the pole. At the third corner is the part of the flagpole that's at the measurer's eye level.

Similar Triangles

If a triangle has two angles that are the same as two angles of another triangle, the two triangles are the same shape. The triangle in the diagram below has two angles that are the same as those of the triangle in the real world, so these two triangles are the same shape.

How do you know that the two triangles have the same angles? People measured an angle with their inclinometers. Then they drew the same angle on the grid paper. The second angle that is the same in both triangles is the 90-degree angle formed by the intersection of the vertical and horizontal lines—that's the vertical line of the flagpole and the horizontal line that marks the measurer's eye level.

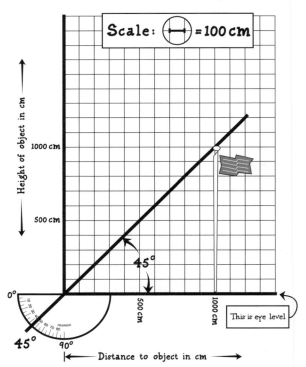

In mathematics, triangles with the same shape are called *similar triangles*. In similar triangles, the sides are *proportional*. That is, if a side of one triangle is half the length of the same side on a similar triangle, then the other sides of the first triangle must be half the length of the corresponding sides of the similar triangle.

Where's the Math?

This activity gives people experience measuring angles and distances. And it uses similar figures to determine distances that would be hard to measure.

Two polygons that are exactly the same shape are called *similar* figures. The angles of a polygon have the same measure as the corresponding angles of a similar polygon. The side lengths of a polygon are proportional to the corresponding side lengths of a similar polygon.

The measures of the three angles of a triangle will equal the measures of the three angles of a similar triangle. But to determine similarity, only two angles are needed. This is because the measures of the three angles of any triangle add to 180 degrees; if the measures of the two angles are equal, the measure of the third angle will also be equal.

Because the sides lengths of similar triangles are proportional, drawing similar triangles and identifying known side lengths can help you figure out lengths you can't measure directly.

Scale drawings such as those done in *Colossal Cartoons* also involve similar figures.

Scaling

The triangle in the diagram on page 146 is a scaled-down version of the one in the real world. How much smaller is the triangle in the diagram? The side of each grid paper square represents 100 centimeters in the world. The squares are about 1 centimeter in length, so the triangle in the real world is about 100 times the size of the triangle in the diagram. By drawing a triangle the exact same shape as the triangle in the real world, people can measure something indirectly that they can't easily measure with a ruler.

When people use their drawings to find the heights of objects, remind them to count the height of each square as 100 centimeters. You should also point out that the horizontal line at the bottom of the grid represents the eye level of the measurer. To find the real height of an object, people need to add the height of their eye level to the height above eye level they got from the grid drawing.

Going Further: Measuring Latitude

On a clear night in the Northern Hemisphere, people can also use their inclinometers to figure out their latitude.

Most globes and maps of Earth are marked with *lines of latitude*—imaginary lines that circle the globe, parallel to the equator. Geographers use latitude to say how far north or south of the equator a place is.

The first step in figuring out latitude is finding the North Star (also known as Polaris). The easiest way to find the North Star is to first locate the Big Dipper. An arrow

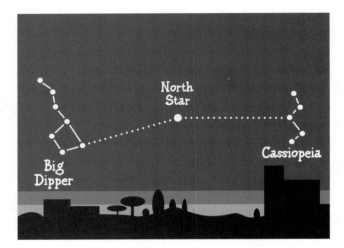

drawn through the two stars that form the end of the Big Dipper's bowl points to the North Star, which is at the end of the handle of the Little Dipper. The North Star is always located between the Big Dipper and the constellation Cassiopeia. In the Northern Hemisphere, these constellations never set.

Like the North Star, they can always be seen on a clear night.

Once people have found the North Star, they sight on it with their inclinometers. The angle of elevation of the North Star, in degrees, is the latitude where they live. The North Pole is at 90 degrees north latitude. Someone standing at the North Pole would have to look straight up to see the North Star.

You might consider having people take their inclinometers home to measure the North Star's angle of elevation. Their answers should all be about the same. You can suggest that if they have a globe or a map at home, they can look up the latitude to check their measurements. You can check as well to confirm their answers. The latitude of the Exploratorium, in San Francisco, California, is about 38 degrees north.

Stomp Rocket!

It's easy to build a simple rocket launcher from a plastic soda bottle—and launching paper rockets is a lot of fun. Experimenting with these simple "stomp rockets" and measuring how high they fly offer practice in measuring angles and experience with similar triangles, a concept many students struggle with.

Before people do this activity, they need to complete *Height Sight* (page 138). In that activity, they build inclinometers, devices that enable them to measure how high their rockets fly. We suggest that you allow one session for *Height Sight* and two for *Stomp Rocket!* That's a lot of time, but *Stomp Rocket!* is a lot of fun.

Preparation and Materials

You will need to find a place to launch the rockets. On a day that's not very windy, a playground or park is a fine launch site. You can also launch rockets in a gym with a high ceiling.

You will need to build rocket launchers, a task that takes 5–10 minutes per launcher. You can build them yourself or have people from your group do it.

A group of 15 people can use one rocket launcher, but we suggest that you make a spare as well. We also recommend having a couple of extra soda bottles handy in case you need replacements.

To build each rocket launcher, you will need:

- 2 empty 2-liter plastic soda bottles from the recycling bin (one for the launcher and one as a spare)

- about 60 cm (2 ft) of PVC pipe with $\frac{1}{2}$-inch inner diameter

- about 1 meter (3 ft) of clear, flexible vinyl tubing with $\frac{1}{2}$-inch inner diameter and $\frac{5}{8}$-inch outer diameter (The

Planning chart

Building rockets	about 30 minutes
Practicing with inclinometer and rocket	about 20 minutes
Launching rockets	about 30 minutes for 20 kids
Figuring out how high the rockets flew	about 20 minutes
Continued experimentation	open-ended

type of tubing doesn't matter, as long as you can tape one end to the neck of the soda bottle and the other end to the PVC pipe.)

- duct tape
- a copy of *Building a Rocket Launcher* (page 151)

To make and launch the rockets, and to calculate how high they fly, you will need:

- a sheet of $8\frac{1}{2}$-by-11-inch paper for each person (Reused paper is fine.)
- scissors (at least 1 pair for every 3 people)
- transparent tape
- markers
- a 3-by-5-inch card (or other stiff paper) for each person
- about 30 cm (1 ft) of PVC pipe for every 5 people (the same type of PVC pipe used for the rocket launcher)
- a meterstick or another way to measure 1 meter
- 10 meters of string
- chalk, masking tape, sticks, or rocks (for marking the launch site)
- rulers (at least 1 for every 3 people)
- pens or pencils
- copies of *Making a Rocket, Rocket Launch Data Sheet, Rocket Height Calculator,* and *Rocket Height Calculator Grid* (1 each per person)

Ask the hardware store to saw the PVC pipe into the lengths you need. If they won't, you'll also need a hacksaw blade; wrap one end of the blade with duct tape to make a "handle," and cut the pipe yourself.

Using This Activity

After an initial session of rocket launching, people often want to improve their rocket designs. More tips on using *Stomp Rocket!* start on page 158 and include suggestions for continued experimentation.

Building a Rocket Launcher

You can make a rocket launcher from an old soda bottle.

What Do I Need?

◇ 2-liter plastic bottle (Only one is required for each launcher, but we strongly recommend that you make two launchers and have a couple of extra soda bottles on hand in case one fails.)

◇ 60 cm (about 2 ft) of PVC pipe with $\frac{1}{2}$-inch inner diameter

◇ 1 meter (about 3 ft) of clear, flexible vinyl tubing with $\frac{1}{2}$-inch inner diameter and $\frac{5}{8}$-inch outer diameter

◇ duct tape

What Do I Do?

Step 1 Remove the cap from the bottle.

Step 2 Insert about an inch of flexible tubing into the bottle opening. Tape it in place with duct tape. Make the connection between the tubing and the bottle airtight. You can test the connection's airtightness by covering the end of the tubing with your hand and squeezing gently on the bottle.

Step 3 Push the PVC pipe against the other end of the flexible tubing. (Don't try to insert the tubing into the PVC pipe.) Tape the tubing and the PVC pipe together. Again, make the connection airtight.

Your finished launcher should look something like this.

Making a Rocket

Follow this design to make a paper rocket, and then test how high it flies.

What Do I Need?

◊ a sheet of $8\frac{1}{2}$-by-11-inch paper

◊ transparent tape

◊ a marker

◊ a 3-by-5-inch card (or stiff paper)

◊ scissors

◊ 30 cm of PVC pipe (One piece can be used by 3 to 5 rocket builders.)

What Do I Do?

Step 1 Roll the sheet of paper into a tube that will fit over the PVC pipe. (You can roll the paper the long way or the short way.) The paper should not be tight around the PVC pipe. You should be able to slide it off *easily*. Tape the paper tube so that it stays rolled up, and slip it off the pipe.

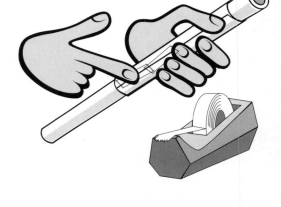

Step 2 Clip the end of the paper tube to make a point. Use tape to seal the point so that it is airtight. This will be the rocket's "nose."

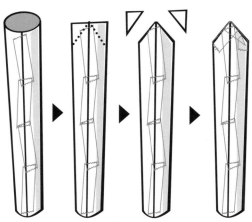

The Math Explorer
Published by Key Curriculum Press / © 2003 Exploratorium

Step 3 Rocket fins will help your rocket fly straight. Fins are usually triangular shapes. Cut fins from a 3-by-5-inch card or some other stiff paper. Here's one way to cut them from a 3-by-5-inch card.

Fold the card in half to make a short rectangle. Unfold the card and cut along the fold line.

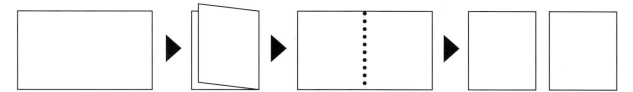

Stack the two halves of the card. Draw a line from one corner of the folded rectangle to the other. Cut along the line you've drawn, and you will have four fins.

Step 4 Tape the fins to the sides of the rocket at the base. Be sure to tape both sides of each fin to the rocket.

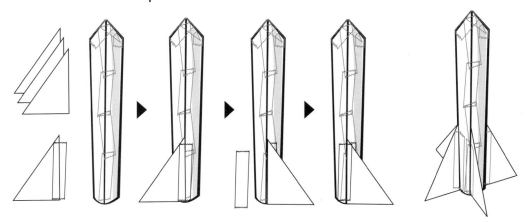

Step 5 Use a marker to write your name on the side of your rocket.

Rocket Launch Data Sheet

Your group leader will help you set up the launch site and launch your rocket. Use your inclinometer to measure how high each person's rocket flies. Use this sheet to keep track of the results.

Name of Rocket Builder	Angle on Inclinometer
1.	degrees
2.	degrees
3.	degrees
4.	degrees
5.	degrees
6.	degrees
7.	degrees
8.	degrees
9.	degrees
10.	degrees
11.	degrees
12.	degrees
13.	degrees
14.	degrees
15.	degrees
16.	degrees
17.	degrees
18.	degrees
19.	degrees
20.	degrees

Rocket Height Calculator

Now you're ready to figure out how high your rocket flew.

What Do I Need?

◊ a copy of *Rocket Height Calculator Grid*

◊ a pencil and a ruler

What Do I Do?

Step 1 Collect measurements for your rocket's flight from three people. Each person will probably get a slightly different measurement. Also ask for each person's eye level height.

Measurement 1: _____ degrees   **Eye level height:** _____ cm

Measurement 2: _____ degrees   **Eye level height:** _____ cm

Measurement 3: _____ degrees   **Eye level height:** _____ cm

Total: _____ $\frac{degrees}{3}$ = _____ average angle in degrees   **Total:** _____ $\frac{cm}{3}$ = _____ average eye level in cm

Step 2 To improve measurement accuracy, scientists sometimes take many measurements and average the results. To get an average, add the three measurements for your rocket and divide by 3. Add the eye level measurements and divide by 3 to get the average eye level.

Step 3 Take a look at the *Rocket Height Calculator Grid*. Each division on the grid represents 100 cm in the real world. Imagine that your eye is right by the protractor in the lower left corner of the grid. The line at the bottom of the grid is a line parallel to the ground, right at the level of your eyes.

Step 4 On the protractor in the corner of the grid, find the average angle that you calculated in Step 2. Mark this angle on the protractor.

Step 5 Place your ruler so you can draw a line through the mark on the protractor and through the corner of the grid. Extend this line all the way across the grid.

Step 6 The people measuring how high your rocket flew stood 10 meters (1000 cm) from where the rocket was launched. That's why the rocket launch site, marked with a rocket on the grid, is 10 squares away from the center of the protractor. Starting at the rocket launch site, draw a straight line that's perpendicular to the eye level line. Extend this line up until it crosses the first line you drew.

Step 7 Now you have a triangle. At one corner of the triangle is your eye, at another corner is the rocket launch site, and at the third corner is the rocket at the top of its flight.

The height of the triangle tells you how high your rocket flew above eye level.

The rocket's height above eye level is _____ cm.

Step 8 Add the average eye level height from Step 2 to the number you calculated in Step 7 to get the approximate height of the rocket's flight.

The height of the rocket's flight is _____ cm (height above eye level plus average eye level).

The Math Explorer
Published by Key Curriculum Press / © 2003 Exploratorium

Rocket Height Calculator Grid

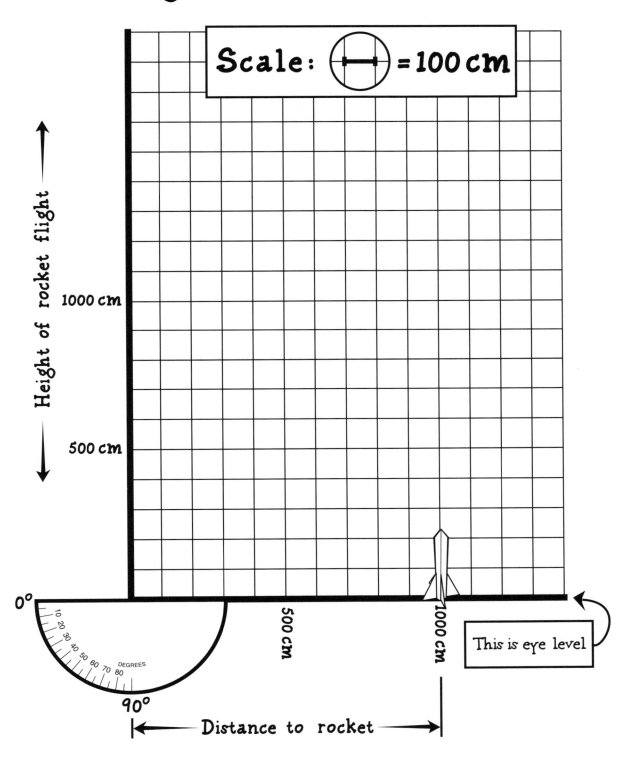

Scale: (⊙—⊙) = 100 cm

Height of rocket flight

1000 cm

500 cm

0°

10 20 30 40 50 60 70 80 DEGREES

90°

500 cm

1000 cm

This is eye level

|← Distance to rocket →|

Using *Stomp Rocket!* with Your Group

Stomp Rocket! combines science and math in an exciting activity. This activity involves making and launching rockets, using an inclinometer to measure each rocket's flight, and figuring out how high each rocket flew.

Your group will discover that math can be useful for judging how well a particular rocket performs. People will learn the value of representing the real world with a diagram, because drawing a diagram lets them calculate the height of the rocket's flight, something that's hard to measure directly.

Introducing the Activity

At the Exploratorium, we introduce this activity by launching a rocket once, inside the room, simply to show the group how cool it is. If you do this, be sure to point the launcher away from the group so that the flying rocket won't hurt anyone or anything.

After launching the rocket, blow into the PVC pipe to reinflate the soda bottle so you are ready to launch again. Do this by putting your hand around the end of the pipe to make a mouthpiece and putting your lips against your hand (rather than the pipe).

Before the rocket building begins, you might check that all members of the group have the inclinometers they made in *Height Sight* (page 138). They'll need them to measure how high their rockets fly.

Making Rockets

Have group members follow the instructions on *Making a Rocket* (page 152), or lead them through the steps yourself.

While people are making rockets, we suggest that you establish a launch order. Have people sign up for launch order only after they have completed a rocket. You might say something like, "Come up and show me your rocket, and I'll sign you up for a launch time."

It's important to establish who is launching when *before* you go to the launch site. (At the launch site, things can get a little chaotic.) Have everyone fill in the names, in order, on their data sheets before going to the launch site.

Before Going to the Launch Site

Here are a few preparations that will make it easier to keep track of what's going on at the launch site. The "Explorer's Notebook" pages do not include instructions for launching, so be sure your group is clear about what to do.

The Math Explorer
Published by Key Curriculum Press / © 2003 Exploratorium

Review How to Use the Inclinometer

If you need help remembering how to use this tool, see pages 140 and 145 of *Height Sight*.

Explain to group members that when they are using their inclinometers to follow a rocket's flight, it's important that they keep both eyes open. This makes it possible to track the rocket in flight. They can either sight through the tube with both eyes open or sight along the top of the tube; both methods will work.

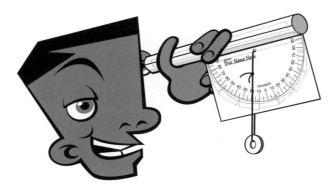

Also, make sure they understand that they will stop tracking the rocket's flight when it reaches its highest point. At that point, they pinch the string against the card and then read the angle off the protractor. Tell them that you will do a couple of trial runs so that everyone gets a chance to practice.

Designate an Aimer and a Launcher

Each time a rocket is launched, you'll need an Aimer and a Launcher. The Launcher stomps the bottle with *one foot* to launch the rocket. (Jumping on the bottle with both feet isn't safe; it's easy to slip on the plastic.) We suggest that you have each person be the Launcher for his or her own rocket.

Aimers are responsible for holding the PVC pipe straight up and making sure the rocket doesn't hit anyone (including themselves). We suggest that you appoint responsible people to be Aimers so that rockets are not aimed or fired at people.

Aimers also reinflate the soda bottle after each launch by blowing through one hand into the PCV pipe.

Explain the Launching Procedure

Tell people that they will follow these steps at the launch site:

- They will mark the launch site.

- They will mark part of a circle that's 10 meters from the launch site. Everyone who is measuring a rocket's height will stand on this arc, so each person is 10 meters from the launch site. (You can mark the launch site and the 10-meter distance from the site before you take your group to the launch site.)

- Each person will launch his or her own rocket with the help of the Aimer.

- Everyone else will measure how high the rocket goes and mark the angle measurement on their data sheets.

Launching the Rockets

Here's What to Take to the Launch Site

Each member of your group needs a *Rocket Launch Data Sheet* on which he or she has written the launch order. Each person also needs a rocket, an inclinometer, and a pen or pencil.

You also need your rocket launchers and a spare bottle or two, some duct tape for emergency repairs, 10 meters of string, and some way to mark the launch site and the circle that's 10 meters from the launch site. You can mark the site with chalk if you're on asphalt, with tape if you're in the gym, or with sticks or rocks if you're on grass.

Getting Ready To Launch

First, mark the launch site. Next, have someone hold one end of the 10-meter string on the launch site while someone else stretches out the other end and marks an arc that is exactly 10 meters from the launch site. Everyone watching the launch will stand on the edge of this arc.

Then do a trial launch. Have everyone count down, chanting together: "3, 2, 1, *Launch!*" On "*Launch!*" the Launcher stomps down on the plastic bottle, sending the rocket flying. Have everyone use his or her inclinometer to measure the angle of the highest point of the rocket's flight.

Have people compare their inclinometer measurements. It will take some practice before they are comfortable using their inclinometers to track a rocket.

Blast Off!

Start launching rockets in the order you previously established. Ask everyone to measure the angle of each rocket's flight and record it on his or her data sheet. If you can, measure the angle yourself—so you can check the others' readings. When we do this activity at the Exploratorium, we stop after each launch and announce our angle measurement.

If you have time, you might want to have everyone launch his or her rocket twice, and then use the best height.

After the group finishes launching their rockets, go back inside to figure out how high the rockets flew.

Figuring Out How High Each Rocket Flew

The procedure here is very similar to the procedure you followed in *Height Sight*.

People will end up with a diagram that looks like the one below. The height of the triangle is how high the rocket flew above eye level. To get the height the rocket flew above the ground, people need to add on the distance from the ground to their eye level. They have collected eye level heights from their measurers and calculated an average. Make sure they add this number to their rocket height.

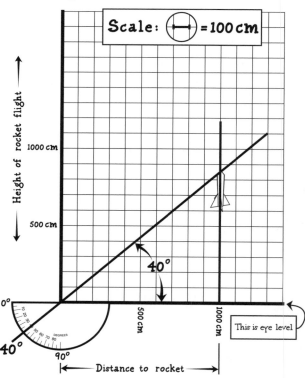

The Math Explorer
Published by Key Curriculum Press / © 2003 Exploratorium

Why Use an Average Angle Measurement?

Some people may ask why they had to average three measurements. People often think of measuring as precise, but it isn't really. Every measurement is an estimate, a best guess at an answer. When people are using a new tool, like an inclinometer, they will come up with different measurements. Averaging these measurements improves the accuracy of the results.

Going Farther: Building a Better Rocket

After people have launched their rockets, they may want to see whether they can build better rockets. They can spend another session improving their rocket designs.

Talk with your group about what changes might make a rocket fly better. Here are some things you can encourage members to think about.

To move through the air, a rocket has to push air aside. Things that travel fast—like sports cars and jets—are shaped to minimize the amount of air they have to push aside to move forward. The fins on the back of a rocket help it slide through the air easily with its nose forward. People might want to experiment with adding more fins, using different-size fins, or removing some fins.

Ask them to imagine that they are pushing a kid on a wagon. With the same push, they can make a little kid in a wagon roll farther than a big kid. The rocket launcher gives your rocket a push with a puff of air. If two rockets have exactly the same design and weight distribution, the lighter rocket will fly farther with the same push than the heavier one. Can they make their rockets lighter?

Ask them to think about other changes they might make. Do long rockets fly higher than short ones? Are some people better Launchers than others? Have people work together to see how high they can make a rocket fly.

Going Farther: Another Experiment to Try

If your group wants to keep experimenting with rockets, here's another suggestion. You'll need a big space (like a playground or a football field).

So far, people have been launching rockets straight up to see how high they will fly. Now ask them to try launching to see how great a *distance* they can get a rocket to fly.

To get the greatest height, people simply point the rocket launcher straight up. To get the greatest distance, they'll need to change the angle at which they hold the launcher. What angle gives them the greatest distance? Encourage them to experiment to find out. By attaching an inclinometer to the launcher's pipe, they can find out at what angle the pipe is held for each launch. If they've done *Stride Ruler* (page 188), they can pace off the distance and figure out how far each rocket flew.

Tetrahedral Kites

This unique-looking kite is fun to build and fun to fly. It's made from four tetrahedrons—three-dimensional shapes with four triangular faces and four corners. (*Tetra* means "four" in Greek.)

Preparation and Materials

We suggest having each person build his or her own kite. For each kite, you will need:

- 24 drinking straws either 8 inches or $7\frac{3}{4}$ inches long

- kite string (about 7 meters for construction, plus at least 20 meters for flying)

- metersticks or rulers for measuring string

- scissors

- one 20-by-26-inch sheet of colored tissue paper (Substitute newspaper for tissue paper if you have to; the Sunday comics make colorful kites.)

- 2 paper clips

- glue stick

- a 1-meter-long strip of crepe paper or newspaper for kite tail (optional)

- copies of *How to Build a Tetrahedral Kite*, Parts 1, 2, and 3, and *Tetrahedral Kite Tissue Pattern*.

Planning chart

Assumes one kite per person	
Part 1: Making the tetrahedrons	about 1 hour
Part 2: Cutting and attaching the tissue paper	about 30 minutes
Part 3: Putting the kite together	about 30 minutes
Flying the kites	open-ended

Using This Activity

We recommend that you build a kite before having your group construct them. There are three reasons for this:

- Doing so will make you a kite-building "expert," and you'll be better prepared to help the people in your group.

- Your finished kite will look so awesome that everyone will want to make one.

- Your kite can serve as a model during the building process.

You should also figure out a time and place to fly your kites once you're done.

For your group, assembling a kite will take one person about two hours. If you are limited to one hour, it's possible for a group of four people to work together to build a single kite. Each person makes one of the four tetrahedrons necessary for one kite. (Be prepared to referee when all four group members want to take the kite home!) People can also work in pairs to bring the combined building and flying time down to less than two hours.

If people all make their own kites, we suggest that you have them work in pairs. It's easier to assemble a kite if one person holds the straws while the other ties the knots.

Tips on flying tetrahedral kites, and more tips on using this activity, start on page 171.

How to Build a Tetrahedral Kite

Follow these instructions to build a cool kite. It's weird looking, but it really flies!

What Do I Need?

◊ 24 drinking straws

◊ a ball of kite string

◊ a meterstick or ruler

◊ a glue stick

◊ scissors

◊ one 20-by-26-inch sheet of colored tissue paper

◊ 2 paper clips

◊ a partner

◊ a copy of *Tetrahedral Kite Tissue Pattern*

◊ a 1-meter-long strip of crepe paper or newspaper for kite tail (optional)

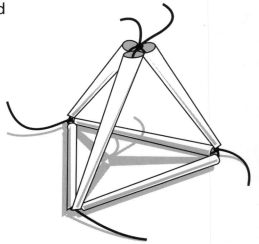

This is a tetrahedron constructed from straws.

PART 1: MAKING THE TETRAHEDRONS

First, you make four strange shapes called tetrahedrons. A tetrahedron is a three-dimensional shape with four sides.

Step 1 Measure and cut 16 pieces of string. You will need these lengths of string:

◊ 4 pieces measuring 100 cm (about 40 in.)

◊ 4 pieces measuring 60 cm (about 24 in.)

◊ 8 pieces measuring 15 cm (about 6 in.)

Step 2 Take one of the 100-cm-long strings and thread three straws onto it.

There are two ways to do this. One is to hold each straw upright and lower the string down through it. Licking the end of the string does not help (it merely makes the string stick to the inside of the straw).

The other way is to suck the string through the straws. Stick a little bit of string into one end of the straw, and then suck through the other end. Hold the other end of the string while you suck, so you don't accidentally swallow the string! You *will* end up with string in your mouth.

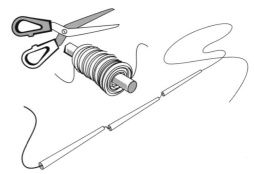

Step 3 Make a triangle with the three straws. Have your partner hold the straws while you tie the two ends of the string in a knot, leaving one end of the string about 8 cm (3 in.) long and the other end very long. Tie this tight so that your triangle is stiff, not floppy.

Step 4 Tie one of the 15-cm-long strings onto one of the corners that has no strings dangling from it. Tie it so that both ends are about the same length.

Step 5 Tie one of the 60-cm-long strings onto the only other corner that doesn't have string hanging from it. Tie it so that one end is about 8 cm (3 in.) long and the other end is very long.

Step 6 Thread two straws onto the long end of the string you just tied on. Then tie the end of the string to the 15-cm piece of string you tied on in Step 4. *Tie it tight!*

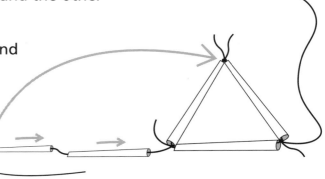

Step 7 You now have a diamond shape. Tie a 15-cm-long string to the corner of the diamond that has no strings dangling from it.

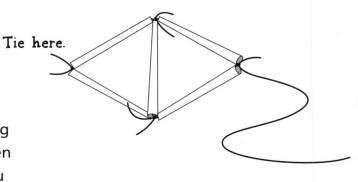

Tie here.

Step 8 Thread one straw onto the long end of the very first string you tied. Then tie that string to the piece of string you added in Step 7.

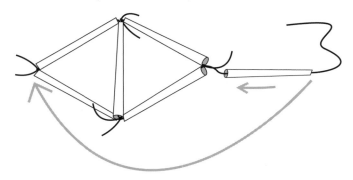

After you tie the last knot, you should have something that looks like this.

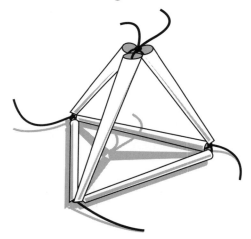

Voila! You've made a tetrahedron!

Step 9 You'll need a total of *four* tetrahedrons to make a kite. So get busy and make three more!

PART 2: CUTTING AND ATTACHING THE TISSUE PAPER

Now that you've made four tetrahedrons, you need to cover two sides of each one with tissue paper.

Step 10 Fold your sheet of tissue paper in quarters. To do this, fold it neatly in half. Then fold it in half again. The new rectangle will be 10 by 13 inches.

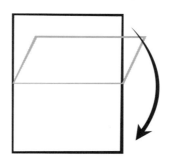

Fold this way . . .

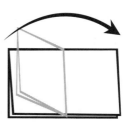

then this way . . .

Step 11 Now fold your rectangle in quarters again. To do this, simply repeat Step 10. The new rectangle will be 5 by $6\frac{1}{2}$ inches.

to get this.

Step 12 Cut out the tissue pattern. If your straws are 8 inches long, use the whole pattern. If your straws are $7\frac{3}{4}$ inches long, cut along the dashed line to make the pattern a little smaller.

Step 13 Put the pattern on top of your folded rectangle of tissue paper. Find the two sides of the pattern that say "Place along folded edge." Line these two sides up with the two sides of your rectangle that have folds—the sides that *don't* have any single-sheet edges. Use paper clips to hold the pattern in place.

Check that you have placed your pattern on the fold the same way everyone else has.

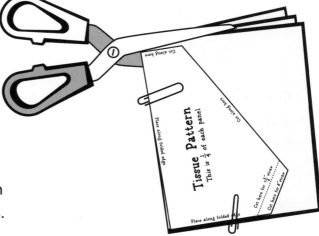

Tissue Pattern
This is $\frac{1}{4}$ of each panel

Place along folded edge

Cut along here

Place along folded edge

Step 14 Cut along the three other sides of the pattern. Then unfold the tissue again. You should have four sheets that look like those to the right.

Step 15 Place any straw of a tetrahedron on the center fold of a piece of tissue paper.

Use a glue stick to put strips of glue on the two flaps of paper lying outside the triangle. Then fold the flaps over the straw, and press them so they stick.

Step 16 Flip the tetrahedron toward the other side of the paper. Again, put a strip of glue on the two flaps of paper that stick out from the triangle. Fold the flaps in, and press them down to make them stick. You've finished putting tissue paper on one tetrahedron.

Step 17 If you want a kite with two or more colors, trade tissue paper with other people. Then glue tissue paper to the other three tetrahedrons by repeating Steps 15 and 16.

Tetrahedral Kite Tissue Pattern

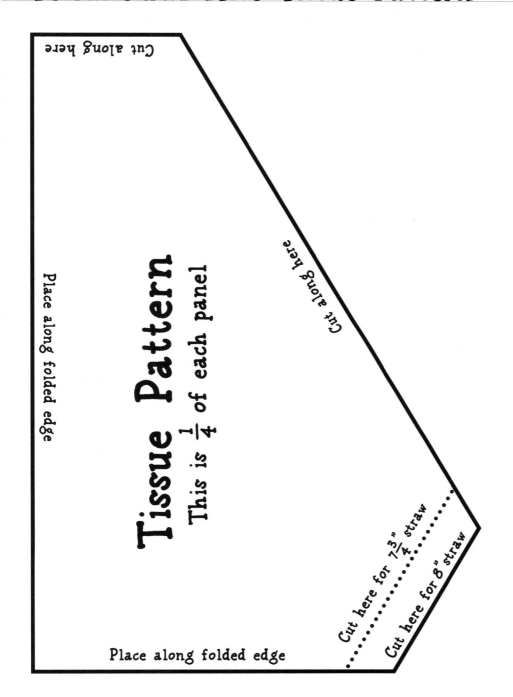

Cut along here

Place along folded edge

Tissue Pattern
This is $\frac{1}{4}$ of each panel

Cut along here

Cut here for $7\frac{3}{4}$" straw

Cut here for 8" straw

Place along folded edge

PART 3: PUTTING THE KITE TOGETHER

You now have four tetrahedrons, each with two tissue-paper sides. Now you will put them together to make a kite.

Step 18 Arrange three of your tetrahedrons in a triangle, with all the paper panels facing in the same direction.

Step 19 Use the pieces of string that hang off each tetrahedron to tie the three tetrahedrons together tightly.

Step 20 The fourth tetrahedron goes on top of the other three. Make sure this one is facing the same way as the others. Have your partner hold the last tetrahedron in place while you tie it to the other three.

Step 21 Trim off any leftover strings. Tie your kite string to the corner shown in the picture. Your kite is ready to fly!

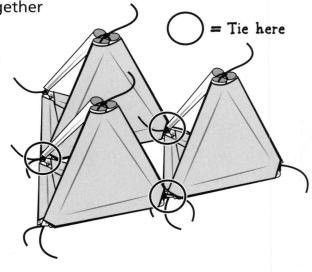

◯ = Tie here

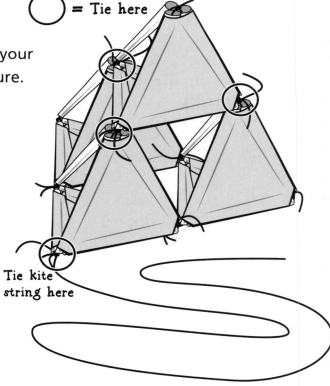

◯ = Tie here

Tie kite string here

Making and Flying Tetrahedral Kites

Making and flying tetrahedral kites with your group will take at least three one-hour sessions, but the end product is well worth the time. When you are done, each person will have a beautiful kite to fly and take home. Making the kites will give group members a chance to practice measuring lengths and a greater familiarity with metric measurements.

If group members have done the *Height Sight* activity (page 138), they can measure the height of their kites in flight, using the inclinometers from that activity.

Helping Your Group Construct Kites

Making the Tetrahedrons

Give each person a copy of *Making the Tetrahedrons* (page 164). We suggest that you demonstrate each step in the building process. Be sure to demonstrate the two methods of threading the straws onto the string.

We also suggest that you have people work in pairs. One person can hold the straws while the other ties the knots. Some members of your group may need help tying their knots. If they are working in pairs, they can manage this without your help.

Many people accidentally let their straws slide off the strings as they are building. They can avoid this problem by always keeping the half-built tetrahedron on a table or other flat surface.

A *tetrahedron* is a three-dimensional shape with 4 triangular faces, 4 corners, and 6 edges. It is the simplest way to enclose space, because no space-enclosing shape with faces (polyhedron) has fewer than four faces. (Ask your group if anyone can think of a way to enclose space using a shape with only three faces.) By contrast, a cube has 6 faces, 8 corners, and 12 edges. A pyramid with a square base has 5 faces, 5 corners, and 8 edges.

R. Buckminster Fuller (1895–1983)

Buckminster Fuller was a well-known engineer, mathematician, philosopher, and inventor. Among his many innovations was a special approach to geometry that he called *synergetic geometry*. In this system, the tetrahedron is the basic unit of space (instead of the cube used in traditional Euclidean geometry). Synergetic geometry is used to understand things like the shapes of viruses and molecules. Fuller is probably best known as the inventor of the *geodesic dome* and as a pioneer in using basic geometric shapes as design elements.

Cutting the Tissue Paper

Give each person a copy of *Cutting and Attaching the Tissue Paper* (page 167) and *Tetrahedral Kite Tissue Pattern* (page 169).

If the straws you are using are $7\frac{3}{4}$ inches long, have everyone cut the short extra strip from the pattern. For 8-inch straws, use the whole pattern.

If the pattern isn't lined up against the proper edges, the tissue-paper shapes that result will not work. Unless you have a large supply of tissue paper, you may want to insist that people have you check that they've lined up their patterns correctly *before* they cut the tissue paper.

Attaching the Tissue Paper

Before people start gluing on the tissue paper, they may want to trade some of their paper pieces with each other to create multicolored kites.

Remind everyone that the paper is fragile and must be handled carefully, especially while applying the glue.

Putting the Kite Together

Give each person a copy of *Putting the Kite Together* (page 170).

Before people assemble their kites, make sure everyone has his or her tetrahedrons pointing in the same direction.

Be sure people tie their kite strings to one of the corners where two tissue-paper panels meet. (See the illustration on page 170.)

People can makes spools for their kite string by wrapping string around a pencil, a stick, or a rectangular piece of cardboard.

Flying Tetrahedral Kites

When and Where to Fly Kites

Fly the kites in a large open area, clear of trees and overhead power lines. Sports fields are ideal. Obviously, you can't be guaranteed a breeze, but in general, the wind is more likely to pick up in the afternoon.

Troubleshooting

When we made these kites, they flew fine without tails. However, no two kites will fly in exactly the same way. If a kite seems to spin or veer too much, try attaching a tail. A strip of crepe paper or newspaper roughly a meter long should work.

Remember that these kites are made of paper; keep them out of the rain and fog and away from water.

Maintaining tension on the string will help the kites fly higher. You and your group can find more tips on flying your kites at www.keymath.com/ME.

About Aerodynamics: Why Does It Fly?

The tissue-paper panels of the kite are very light, but they do block wind. When wind hits the panels, it gets deflected *down*. This pushes the kite up. Physics says that every action has an equal and opposite reaction. In this case, the action of the wind moving *down* is balanced by the kite moving *up*.

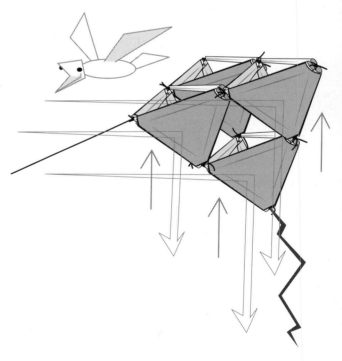

Where's the Math?

If your group has done *Paper Engineering* (page 119), you might ask if this kite reminds people of that activity. Like the first sculpture in *Paper Engineering*, this kite is constructed of small shapes that look just like the large shape.

Each kite is made of four small tetrahedrons that are joined together to make a large tetrahedron. Point this out to members of your group, and help them see the large tetrahedron.

Ask group members how they might make a larger version of the tetrahedral kite. You can make a larger version by starting out with longer straws. By doing

Notice that the big triangle is made up of three smaller triangles. Your kite is a large tetrahedron made up of four smaller tetrahedrons.

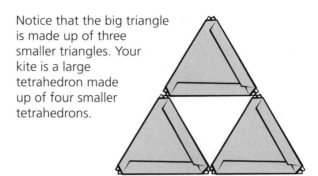

this, you will make each tetrahedron larger, so that the final kite is also larger. Or you can tie together four of your tetrahedral kites to make a giant tetrahedral kite! Try it and see if it will fly.

Flying Things

Making and testing paper airplanes is great fun—and will burn off some energy on a rainy day. In this activity, math is used to make a paper airplane contest fair and to determine which airplane really flew the best.

Preparation and Materials

For each person, you will need:

- a few sheets of $8\frac{1}{2}$-by-11-inch paper (Reused is fine.)
- small paper clips, a pencil, and a ruler
- copies of *Folding Your Flying Thing*, *Testing Your Flying Thing*, and *Flying Things Data Sheet*
- 1-cm graph paper or a copy of *Centimeter Grid Paper* (page 204)

For each pair of people, you will also need:

- a piece of string about 150 cm long
- a meterstick or a *Make-It-Yourself Meterstick* (page 199)

To set up the testing ground where you will fly the planes, you will need:

- masking tape and a permanent marker (if your testing ground is indoors) or chalk (if it is outdoors)
- a meterstick or a *Make-It-Yourself Meterstick* (page 199)

Use masking tape or chalk to mark the ground in 50-cm increments. Label each increment: zero cm (start line), 50 cm, 100 cm, and so on. Your testing ground should be at least 10 meters (1000 cm) long.

Using This Activity

Tips for how to use *Flying Things* start on page 182.

Planning chart

Folding flying things	10 minutes
Testing flying things	15–20 minutes
Calculating and comparing glide ratios	15 minutes
Modifying planes and improving glide ratios	open-ended
Testing planes from various heights	open-ended

The Math Explorer
Published by Key Curriculum Press / © 2003 Exploratorium

Folding Your Flying Thing

In the past two decades, paper airplane makers have introduced some improvements in paper airplane design. This paper airplane includes the *Nakamura lock,* which is named after the origami artist who invented it.

What Do I Need?

◊ a few sheets of $8\frac{1}{2}$-by-11 inch paper

◊ a pencil

What Do I Do?

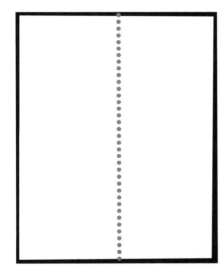 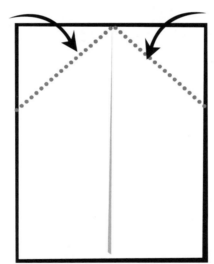 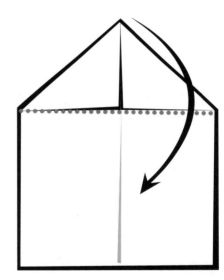

Step 1 Fold a sheet of paper in half lengthwise. Unfold it so that the crease makes a valley in the paper.

Step 2 Fold the top corners down to the center fold.

Step 3 Fold the tip down.

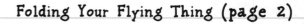

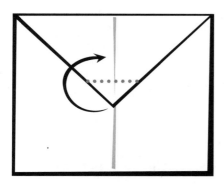

Step 4 Fold about 1 inch of the tip up, and then unfold it.

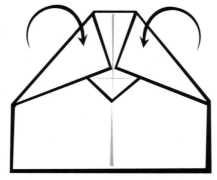

Step 5 Fold the top corners down to the center fold so that the corners meet above the fold in the tip. The top—the nose of the plane—should be blunt.

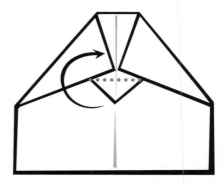

Step 6 Fold the tip up. This is the Nakamura lock.

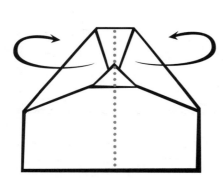

Step 7 Fold the entire plane in half so that the lock is on the outside.

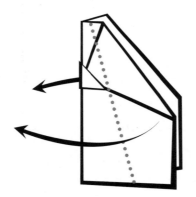

Step 8 Fold the wings down. You can choose how wide or narrow to make the wings.

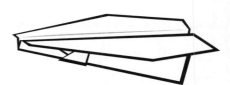

Step 9 Write your name on your plane.

The Math Explorer
Published by Key Curriculum Press / © 2003 Exploratorium

Testing Your Flying Thing

Tall people usually have an advantage in flying paper airplanes: they launch their planes from a greater height. To make this contest a little more fair, you won't just measure how far your plane flies. You're going to calculate your plane's *glide ratio*—the horizontal distance the plane flew divided by the launch height. The plane with the best glide ratio wins!

For safety's sake, follow these rules: Never throw your plane at anyone, and never throw your plane when anyone is in the way.

What Do I Need?

◇ a paper airplane

◇ a pencil

◇ small paper clips

◇ *Flying Things Data Sheet*

◇ a piece of string and a meterstick

◇ a partner

What Do I Do?

Step 1 Have your partner measure the distance from the ground to the top of your shoulder. Use the string to measure the distance. Use the meterstick to measure the string.

Step 2 This distance is your *launch height,* because you'll throw your plane from about shoulder height. Write it on your *Flying Things Data Sheet.*

Step 3 Take your plane and your *Flying Things Data Sheet* to the testing ground with the rest of the group.

Step 4 When your leader says it's time, give your plane a gentle toss forward. Your goal is to have it glide smoothly and gently to the ground. To accurately measure your plane's glide ratio, you have to throw the plane so that it never rises above your shoulder level. Experiment with your throwing technique—sometimes a plane will actually fly a shorter distance if you throw it harder.

Step 5 If your plane doesn't fly well, make a few adjustments. This is known as *trimming* your plane. Here are some adjustments to try:

◆ If the plane dives into the ground, bend up the backs of the wings. A little bend goes a long way.

◆ If the nose of the plane rises first and then drops, the plane is stalling. Bend down the backs of the wings. Keep your adjustments small.

◆ If the nose is still rising, add a paper clip to the nose.

Trim your plane, and practice throwing it until you're happy with how it flies.

Step 6 Your leader will tell you when it's time to test your plane. When it's your turn, throw your plane. Note where the nose of your plane lands, and mark that measurement on your *Flying Things Data Sheet*. If your plane lands between two marks, use a meterstick to measure how far the plane flew past the first mark.

Step 7 Test your plane three times. If you have time, do more trials. On your *Flying Things Data Sheet,* record how far your plane flew each time.

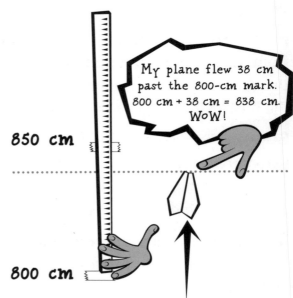

850 cm

My plane flew 38 cm past the 800-cm mark. 800 cm + 38 cm = 838 cm. WoW!

800 cm

The Math Explorer
Published by Key Curriculum Press / © 2003 Exploratorium

Explorer's Name: _____   Date: _____

Flying Things Data Sheet

Use this data sheet to keep track of how well your plane flies.

What Do I Need?

◇ a pencil

◇ a ruler

◇ a sheet of grid paper

◇ a calculator

A ratio is a way of comparing two numbers.

$$D \div H = G$$

Trial Number	Distance Flown (D) (cm)	Launch Height (H) (cm)	Glide Ratio (G)
1			
2			
3			
4			
5			
6			
Average			

What Do I Do?

Step 1 For each trial, divide the distance your plane flew by your launch height to get the glide ratio. Round your answer to the nearest tenth. Write the result—your glide ratio—in the chart above.

Step 2 Figure out your average distance by adding the distance from all your trials and dividing the result by the number of trials. Figure out your *average glide ratio* in the same way.

Step 3 Assume that the side of each square on the grid represents 50 cm in the real world.

Draw a mark on the vertical side of the grid to show your launch height. Draw a mark on the horizontal side of your grid to show the average distance your plane flew. Connect these two marks to make a right triangle (a triangle with a 90-degree angle). The height of the triangle is your launch height. The base of the triangle is the average distance of your plane's flight. The hypotenuse, the longest side of the triangle, shows the approximate flight path of your plane.

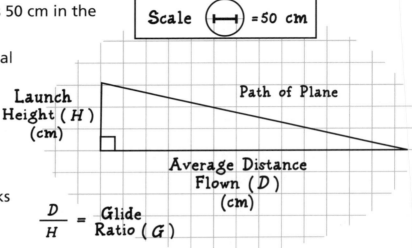

$$\frac{D}{H} = \text{Glide Ratio} (G)$$

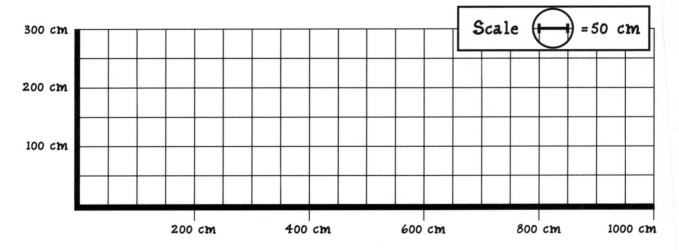

Other Experiments to Try

Change the Launch Height

What do you think will happen if you stand on a chair and throw your plane? What if you stand on something even taller than a chair? Experiment to find out. To calculate the glide ratio for each flight, you'll need to measure your new launch height. That's the distance from the ground (not the surface you're standing on) to the top of your shoulder in each situation.

Standardize Your Throwing Technique

The speed of a plane and the direction and speed of the wind both affect glide ratio. When you fly your paper airplane, changes that may seem small can have a big effect on glide ratio—which is why your glide ratio isn't the same for every flight. One thing that can make a big difference is how you throw the plane. Can you think of a way to standardize how you launch your plane?

Change Your Design

Modify your plane to improve its glide ratio. You might create several different paper airplanes and compare their glide ratios.

Experimenting with *Flying Things*

In this activity, members of your group fold paper airplanes, fly their planes, and compare the planes' performances.

The greater the height from which a plane is launched, the farther the plane has to drop before it hits the ground. A fair comparison of one plane's performance against another's has to take into account the height from which the plane was launched. One way to do this is to calculate the plane's glide ratio.

The *glide ratio* is the distance a plane flew divided by the height from which it was launched. Comparing glide ratios eliminates the advantage a tall person gets by launching a plane from higher above the ground, and makes it possible for people of various heights to compete equally.

Making *Flying Things*

Folding the Paper Airplanes

You can give everyone in your group a copy of *Folding Your Flying Thing* (page 175) and have them read the instructions. Or you can lead the group through the steps as members follow along. Some people may be unfamiliar with the word *origami*. Be sure to define it for them when the instructions in *Folding Your Flying Thing* refer to the Nakamura lock.

We suggest that you have everyone fold the same type of paper airplane. Although some people may want to create planes of a different design altogether, we recommend that you ask them to try this design first.

Some Simple Rules

Things can get a little chaotic when people start testing their planes. Be sure to review these simple rules before they reach this point:

- Never throw your plane at anyone.
- Never throw your plane when anyone is in the way.

We suggest that you have people wait until they get to the testing ground before they start throwing their planes. And we suggest that you have people pair up and measure launch height before they go to the testing ground.

At the Testing Ground

Have everyone take a few practice throws. You may want people to line up and do this one by one. They may need to make some adjustments to their planes before the planes glide smoothly. You'll find suggestions for adjustments in *Testing Your Flying Thing* (page 177). Have people try the adjustments one by one, making only one change before each test flight. Sometimes a paper clip on the nose or a slight adjustment to a wing can make a big difference.

Tell people that throwing a plane harder doesn't always make it fly farther. Give them time to experiment with throwing techniques.

The goal is to throw the plane at the speed that makes it glide the farthest *without ever rising above shoulder height.* To get an accurate glide ratio, the launch height must be the highest point in the plane's flight.

Collecting Data

Have people line up and test their planes one by one. You may need to remind them of the rules: don't throw your plane at anyone or when anyone is in the way. Some people may need help measuring distance when their plane lands between two marks.

Have each person fly his or her plane at least three times and record the results on their *Flying Things Data Sheet* (page 179).

Calculating the Glide Ratio

What's a Ratio?

Many members of your group may have heard the word *ratio* before. The concept of ratio is introduced in middle school. Basically, a ratio is a way of comparing two numbers.

Figuring Out the Glide Ratio

To find the glide ratios for their trial flights, members of your group should divide the distance a plane flew by the height from which that plane was launched. Rather than writing out such terms as *distance flown* or *launch height*, scientists and mathematicians assign letters to represent each value.

Here, *D* represents the distance the plane flew, *H* represents the launch height, and *G* represents the glide ratio. Using these letters, here's the equation for calculating glide ratio:

$$\frac{D}{H} = G$$

The greater *G* is, the better a flying thing glides!

Rounding Up and Rounding Down

If your group uses calculators to get the glide ratios, the answers may have a long string of numbers to the right of the decimal point.

They may ask how many of these numbers they should write down.

Calculators are very precise—much more precise than the measurements made in this activity. Being very precise in a calculation when the measurements are not so precise doesn't make sense. So you may want to suggest that they round their answers to the nearest tenth, keeping just one digit to the right of the decimal place.

The concept of *rounding* is introduced in elementary school, but many middle school students still struggle with it. If you want to walk your group through the process of rounding, have them look at their answers and identify the tenths place and the hundredths place.

To round to the nearest tenth, they need to look at the hundredths place to figure out what to do. If the digit in the hundredths place is less than 5, then leave the digit in the tenths place unchanged and drop the digits to the right. If it is 5 or more, they add 1 to the digit in the tenths place and drop the digits to the right of the tenths place.

Suppose the number is 5.632. Because the digit in the hundredths place is 3, the number is rounded to 5.6.

Suppose the number is 5.673. Because the digit in the hundredths place is 7, the digit in the tenths place is increased by 1, and the number is rounded to 5.7.

Drawing a Diagram

Many people find it easier to understand mathematical concepts when they can draw a picture or a diagram. Tell members of your group that they will draw a picture "to scale" to represent their plane's average path. A scale drawing looks just like the original, but in a different size.

To make a drawing to scale, they need to figure out what distance a side of a square on the grid represents in the real world. Each square on their grid paper represents 50 centimeters in the real world. Even people whose planes flew 1000 centimeters can fit their scale drawings on the grid.

By comparing diagrams, your group can compare the flight paths of different paper airplanes. The greater the glide ratio, the less steep the slant of the flight path. By looking at the diagrams, people can see that a greater glide ratio means that a plane glides a long way while dropping just a little.

To put the glide ratios of the paper airplanes in perspective, you might tell your group that the average light plane has a glide ratio of about 10 to 1. That means a plane that has no engine can glide 10 meters forward for every meter it drops. So, if the plane is 100 meters above the ground, it can fly 1000 meters horizontally before it touches down. A modern glider—a plane designed to glide—may have a maximum glide ratio of 55 to 1.

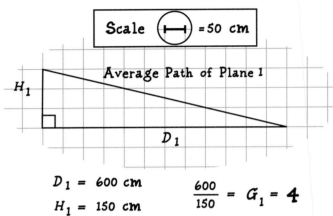

Scale $\vdash\!\!-\!\!\dashv$ = 50 cm

Average Path of Plane 1

H_1

D_1

$D_1 = 600$ cm

$H_1 = 150$ cm

$$\frac{600}{150} = G_1 = 4$$

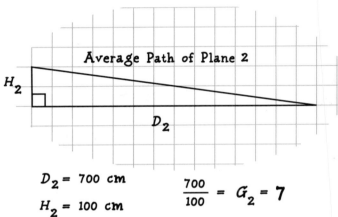

Average Path of Plane 2

H_2

D_2

$D_2 = 700$ cm

$H_2 = 100$ cm

$$\frac{700}{100} = G_2 = 7$$

The Math Explorer
Published by Key Curriculum Press / © 2003 Exploratorium

Other Experiments to Try—Varying Launch Height

It can be fun to launch planes from different heights and measure how far they fly. Ask people to predict how far they think their planes will go when they increase the launch height.

People might want to test whether the glide ratio remains the same for a plane launched from different heights. Have them gather data by launching a plane several times from each of several heights and calculating the glide ratio for each flight. Remember that the *glide ratio* is the flight distance divided by the launch height. If the glide ratio is the same for every height, then doubling the height means the distance will double, too. Is that true for your data?

Chances are it won't be. Tiny changes—a difference in throwing technique, a sudden breeze, or a bend in a wing from the last crash landing—can change a plane's glide ratio.

You can use this opportunity to talk about the fact that scientists conduct tests under controlled conditions. Scientists call the factors that affect an experiment *variables*. In testing their planes, your group can control some variables, but others are more difficult to control.

Other Experiments to Try—Standardizing Throwing Technique

If people want to keep experimenting, encourage them to experiment with throwing technique or airplane design. Warn them that they should make changes one at a time—and test after each change. If someone makes lots of changes to a plane and then tests it, there's no way to tell which changes helped—and which ones hurt—the plane's performance.

Where's the Math?

Making and testing paper airplanes introduces people to one way math can be used: to measure and quantify performance.

Drawing diagrams may also help people see that there are different ways of presenting the same information—and that different ways of looking at information help people understand it in different ways.

Finally, this activity involves work with ratios, an important concept in middle school math.

Tools You Can Make and Use

These materials will help you and your group do the other activities.

Activity 21 **Stride Ruler**

This activity introduces people to a tool they can use to estimate distances: their own two feet.

Activity 22 **Paper Dice**

People can use this template to fold their own dice from paper. These paper dice can be used to play Boxed In! *and* Pig.

Activity 23 **Make-It-Yourself Meterstick**

With this activity, people can make the metersticks they need for various other activities.

Activity 24 **Centimeter Ruler**

People can use copies of this page on card stock to make their own centimeter rulers.

Grid Paper

Copy this page whenever you need graph paper for your group.

A C T I V I T Y **21**

Stride Ruler

After doing this activity, people will be able to estimate distances using their own two feet. This tool comes in handy for *Height Sight* (page 138). *Stride Ruler* introduces the importance of estimates and offers practice in measuring, an important skill for middle schoolers.

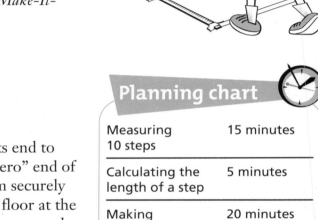

Preparation and Materials

For the entire group, you will need:

- 4 metersticks (you can make your own with *Make-It-Yourself Meterstick*, page 199)

- tape (any kind)

- calculators (1 per 3 or 4 people)

For each person, you will need:

- a copy of *Stride Ruler*

Before beginning this activity, lay the metersticks end to end in a straight line on the floor, placing the "zero" end of one against the "100" end of the next. Tape them securely to the floor, and then tape a "Start Line" on the floor at the zero-cm mark on the first meterstick. This is where people will line up their toes before starting forward.

Planning chart

Measuring 10 steps	15 minutes
Calculating the length of a step	5 minutes
Making *Make-It-Yourself Metersticks* (optional)	20 minutes

Using This Activity

Tips for using *Stride Ruler* start on page 190. If your group does *Stride Ruler* just before *Height Sight* (page 138), members will immediately see why this is a useful way to measure a distance.

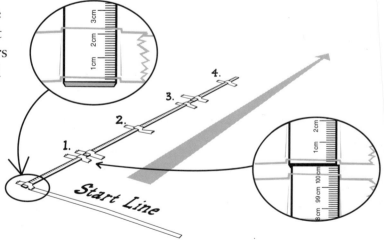

Stride Ruler

You can measure a distance using only your own two feet.

What Do I Do?

Step 1 Put your toes just behind the "Start Line." Take 10 "baby steps" forward, walking right next to the line of metersticks. (In "baby steps," you walk heel to toe, so that the heel of one foot touches the toes of the other foot.)

Step 2 After your tenth step, stop and look at where the toes of your forward foot are. What's the measurement in centimeters (cm) on the meterstick? Write that number down.

Position of my toes after 10 baby steps = _____ cm

Step 3 Now look to see how many metersticks you walked past. For every meterstick you walked past, add 100 cm to the measurement in Step 2. Write down your answer in centimeters.

Length of 10 baby steps in centimeters = _____ cm

Step 4 To find the length of just one step, divide your total measurement by 10.

$$\frac{\textbf{Length of 10 steps in cm}}{\textbf{10}} = \_\_\_\_\_ \textbf{ cm = Length of 1 step in centimeters}$$

Step 5 Now you have a tool that will help you estimate the distance between any two points. Simply walk in baby steps and count your steps. Then use this formula:

Number of baby steps × Length of baby steps = Distance in centimeters

Helping Your Group with *Stride Ruler*

Have each person follow the instructions in *Stride Ruler* on page 189 to measure the length of 10 steps and then compute the average length of 1 step.

Measuring Steps

You may have to show people what "baby steps" are. Simply walk heel to toe.

If possible, help each member of the group read his or her measurement from the metersticks and count how many metersticks were walked past. People need to add 100 cm to their measurement for each meterstick they walked past. Sometimes people forget that *each* meter is 100 cm.

Some people may want to use a calculator to divide by 10. Others may realize that dividing by 10 is easy: just move the decimal point one place to the left. Let group members make this discovery on their own.

Why Measure Ten Steps?

You might ask people if they know why they went to the trouble of measuring 10 steps if all they really want to know is the length of 1 step.

There are two good reasons. First, no measurement is exact; there's always some error. In a sense, every measurement you make is an estimate, no matter how carefully you measure.

Second, the length of each baby step you take will be slightly different. By measuring the length of 10 steps and dividing by 10, your longer and shorter steps balance out to give you the length of an average baby step. Because you have measured 10 steps, you divide any measurement error by 10, giving you a more accurate measurement.

After Completing *Stride Ruler*

Explain to group members that they now have a tool that will help them estimate distance. By counting the number of baby steps from one point to another, they can make a reasonable estimate of the distance between the points.

To give it a try, ask everyone in the group to count how many baby steps it takes to walk a certain distance—the length of the room, for example. Chances are everyone will get a different number of steps.

Convert Baby Steps to Centimeters

Now have people convert their measurements in baby steps to measurements in centimeters. They can do this by following the formula in Step 5 of the *Stride Ruler* instructions. Or you can introduce them to *dimensional analysis*, which is a simple method for figuring out how to convert from one measurement system to another, without memorizing any formulas. See *Where's the Math?* on page 193 for details.

The Math Explorer
Published by Key Curriculum Press / © 2003 Exploratorium

Conversion Challenges

After group members have successfully used baby steps to measure a distance, you might challenge them to convert from centimeters to other systems of measurement.

Convert Centimeters to Inches

Ask people to convert the average length of their baby step in centimeters into inches. Tell them that there are 2.54 cm per inch, or 2.54 cm/in.

Some people won't know how to convert centimeters to inches. They may know that they have to either divide by 2.54 cm/in. or multiply by it, but not know which.

Here's an easy way to help them figure it out. Have them take a look at a ruler marked in both centimeters and inches. Ask, "Will there be more inches than there were centimeters, or fewer?" Most people will realize that there will be fewer inches than centimeters because an inch is longer than a centimeter. Therefore, when centimeters are converted to inches, the answer will be a smaller number.

Now ask, "If you multiply by 2.54 cm/in., do you get a larger number or a smaller number?" Multiplying by 2.54 cm/in. results in a larger number, which they have just told you is wrong. That means the right choice is to divide by 2.54 cm/in.

To convert centimeters to inches, divide the length in centimeters by 2.54 cm/in.

People can use a calculator to do the calculation—and then check their answer by marking out their step in centimeters, and measuring it in inches.

Convert Steps to Miles

Ask people to figure out how many baby steps they would have to take to walk a mile (mi).

Explain that there are

63,360 in./mi   **160,934 cm/mi**

5280 ft/mi

You may have to lead your group through this calculation. You can use the tool of *dimensional analysis*. (See *Where's the Math?* on page 193.) Or follow these steps:

- Take the number of centimeters per mile—160,934 cm/mi.

- Figure out exactly what you want the answer to be, in terms of which unit is where. You want to know how many baby steps there are in a mile. So, you want an answer that is in baby steps per mile, or baby steps/mi. That means you want "baby steps" to be on the top in your answer (in the *numerator*) and "miles" to be on the bottom (in the *denominator*).

- Figure out how you can write down what you already know to put baby steps in the numerator and miles in the denominator:

$$\frac{1 \text{ baby step}}{20 \text{ cm}} \times \frac{160,934 \text{ cm}}{1 \text{ mi}}$$

- Multiply these numbers as you would multiply any fractions (multiplying the numerator by the numerator and the denominator by the denominator). Because you have centimeters in both the numerator and the denominator, this unit cancels out, leaving you with:

$$\frac{160,934 \text{ baby steps}}{20 \text{ mi}}$$

- Divide 160,934 by 20, and you get 8046.7 baby steps/mi.

- Round the answer to 8047 baby steps/mi.

When converting from one unit of measurement to another, it's always important to pay attention to the units of measurement. Including the units provides a way to double-check an answer.

Estimation Challenges

Once people in your group know the length of their stride, they can use that number to make other estimates. Here are some things they can try:

- They can use their stride rulers to measure the distance between two points in the room (or outside). Have them compare results.

- As a research project, they can figure out the distance they walk to school, between classes, or to the store.

Other Handy Ways to Estimate

Tell people in your group that a centimeter is about the width of their little finger. Ask them if they can suggest any other ways to use their bodies or common objects to measure distance.

There are many possible answers to this question. Here are just a few:

- A jumbo paper clip is about 1 centimeter wide.

- A dollar bill is about 6 inches long.

- A dime is just a little more than 1 millimeter thick.

- This page is about 30 centimeters long.

1 mm   1 cm   6 in

Where's the Math?

At school, students may get the idea that math and science are always very precise. In math class, there is usually a precise answer. But in everyday life, you often need only an estimate. You might describe an estimate as a well-informed guess—an answer that may not be exactly right, but is close enough for your purposes.

Here's an example you might share with your group. Suppose you decide to go to the movies Saturday night. You want to figure out how much money you'll need for the movie and a drink and some popcorn. You don't need to know the amount down to the penny—but you do

need to know *about* how much money to take with you. Is 5 dollars too little? Is 20 dollars too much? When you figure out how much money you need for the movie, you are making an estimate.

Many mathematical problems require precise answers, but even in math class, estimation can help a student. It's always useful to estimate an answer before doing the precise calculation. Then a student can check his or her answer against the estimate. If the two are very different, the student knows to double-check his or her calculations.

The Math Explorer
Published by Key Curriculum Press / © 2003 Exploratorium

Where's the Math?

You can show members of your group a tool that will help them figure out how to convert from one measurement system to another, without memorizing any formulas. This tool is known as *dimensional analysis*.

Converting from one unit of measurement (such as baby steps) to another (such as centimeters) requires a *conversion factor*. In the *Stride Ruler* activity, people figured out a conversion factor for baby steps to centimeters. Their conversion factor is the number they calculated in Step 4: the length of one baby step in centimeters.

Ask group members how they might write that number as a fraction. Here are two ways:

$$\frac{20 \text{ cm}}{1 \text{ baby step}} \qquad \frac{1 \text{ baby step}}{20 \text{ cm}}$$

Both ways are correct. You might want to point out that these fractions are both equal to 1. (Remember, any number divided by itself equals 1. So, a fraction in which the top and bottom are equal—whether it's $\frac{2}{2}$, or $\frac{3}{3}$, or $\frac{20 \text{ cm}}{1 \text{ baby step}}$—is equal to 1.)

Ask your group what happens when you multiply a number by 1. (You get the number you started with.) So, multiplying by one of these conversion factors won't change the number—but it will change the units!

The key to converting from one unit of measurement to another is this: pay attention to the units! (That may sound obvious, but lots of people don't do it!) When you multiply fractions, the units act just as numbers do. So, if you have the same unit on the top of one fraction and on the bottom of the other, they cancel out.

Suppose the length of the room is 25 baby steps when measured by someone whose baby step is 20 cm long.

$$\frac{20 \text{ cm}}{1 \text{ baby step}} \times 25 \text{ baby steps} =$$

$$\frac{20 \text{ cm}}{1 \text{ baby step}} \times \frac{25 \text{ baby steps}}{1} =$$

$$\frac{20 \text{ cm} \times 25 \text{ baby steps}}{1 \text{ baby step}}$$

The baby steps cancel out, and the result is

$$\frac{20 \text{ cm} \times 25}{1} = (20 \times 25) \text{ cm} = 500 \text{ cm}$$

What will happen if you use the other fraction, $\frac{1 \text{ baby step}}{20 \text{ cm}}$?

$$\frac{1 \text{ baby step}}{20 \text{ cm}} \times 25 \text{ baby steps} =$$

$$\frac{1 \text{ baby step} \times 25 \text{ baby steps}}{20 \text{ cm}}$$

Look at the units. You'll see that they don't cancel out, so this can't be the right answer!

ACTIVITY **22**

Paper Dice

Your group can use paper dice to play *Pig* (page 14) or *Boxed In!* (page 2). Making these dice requires patience, as well as careful cutting and folding.

Preparation and Materials

To do this project, you will need:

- a pair of scissors for each person
- copies of *Making Paper Dice* and *Paper Dice Template*
- transparent tape (optional)

We suggest that you make one paper die yourself before having your group try it. The wrapping process is tricky, and you will want to be able to demonstrate it.

Using This Activity

You can tell group members that they will be using these dice to play games. *Pig* and *Boxed In!* require one pair of dice per group playing.

Making paper dice requires patience and perseverance, traits that are useful in any learning endeavor. People experienced with origami or other paper-folding activities will have an advantage.

You may want to warn your group at the start that the final steps of making paper dice can be a little tricky. People need to hold the paper cube together while sliding a paper strip through a small opening. You might suggest that they work in pairs when they get to Step 6. One person can hold the cube together while the other slides the strip through.

If people get frustrated at this stage, suggest that they use one small piece of tape to hold Strip A in a cube shape and another to hold Strip B in place. The tape will make the dice a little less evenly weighted but will make the assembly process much easier.

Tips for testing paper dice are on page 198.

Planning chart

Making paper dice	15 minutes
Testing the dice	10 minutes

The Math Explorer
Published by Key Curriculum Press / © 2003 Exploratorium

Making Paper Dice

With careful cutting and folding, you can make
a pair of dice out of paper.

What Do I Need?

◇ scissors

◇ *Paper Dice Template*

◇ clear tape (optional)

What Do I Do?

Step 1 Cut the *Paper Dice Template* in half, and give one
half to a friend. Then cut out the three strips on your half
of the template. Be sure to cut right on the outside lines.
That will make assembling the dice easier.

Step 2 Fold each strip along the solid lines. Make
a sharp crease on each line. Always fold in the same
direction so that the strip of paper forms a box with
the dots on the outside.

Step 3 Take Strip A. Arrange it so that Square A
overlaps the blank square and the strip makes a box
shape with two missing sides. Do the same with
Strip B.

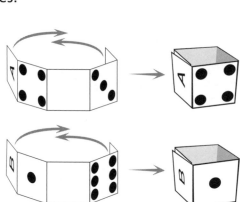

Step 4 Position Strip B so that the square
with six spots is on top. Position Strip A so that
the square with A is on top. Insert Strip A into
Strip B to make a cube with six sides. Make sure
Square B is visible on one side of the cube.

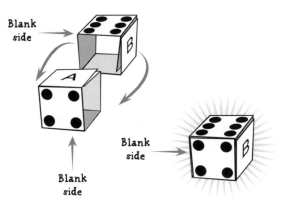

Blank
side

Blank
side

Blank
side

Step 5 Take Strip C. Fold along the dotted lines on Square C and Square D so that the strip is pointed at both ends.

Step 6 Here's the tricky part. Find the edge of your cube where the side with three dots meets Side B. Take Square D, and slide it through the gap at that edge. Slide Square D under the side with three dots, and pull it through from the other side. It'll take some careful maneuvering, but you can do it.

It can be hard to hold your cube together while doing all this sliding. If you can't manage it, take a tiny piece of tape and tape Square B to the square with six dots. Then try again.

Pull Square D through until the square with five dots is right on top of Square B.

Step 7 Tuck Square C under the square with four dots. Then tuck Square D under the same square.

You're done!

The Math Explorer
Published by Key Curriculum Press / © 2003 Exploratorium

Paper Dice Template

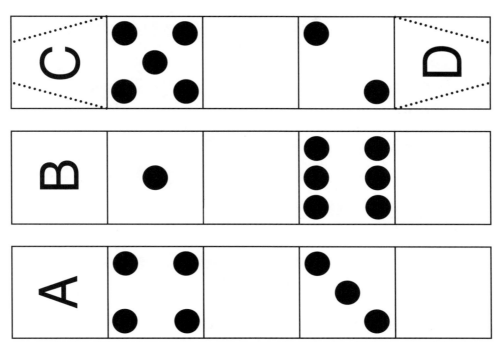

Cut

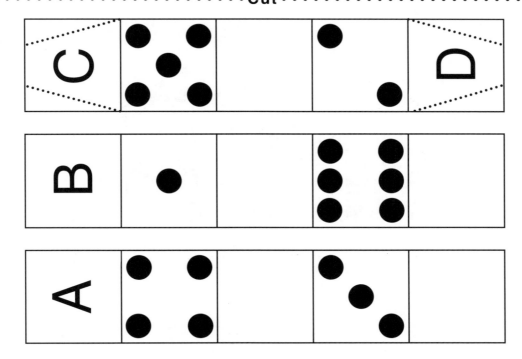

Testing *Paper Dice*

After your group has made paper dice, you can have members test each die to see if it is fair.

Testing the Dice

Ask your group how many faces each die has. (Each die is a cube, and a cube has six faces.) A die that is *fair* is equally likely to land with any of the six faces up—so it is equally likely to roll any number from 1 to 6.

Have people roll each die 100 times and keep track of how many times each number comes up. There will be some variability, but the number of times each number comes up should be more or less equal.

If it isn't, ask people how they might suggest making a die more fair. If one number comes up more often than all the others, you could weight the die by putting a little bit of clear tape on that number, making it heavier and less likely to come up.

Where's the Math?

When group members test their dice, they are experimenting with probability. To learn more about probability, see page 14.

Make-It-Yourself Meterstick

If you don't have metersticks to use with your group, you can have members make their own. Once they have made these, they can use them for a variety of measurement activities.

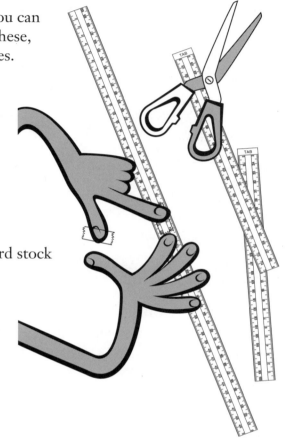

Preparation and Materials

To do this project, you will need:

- a pair of scissors for each person
- transparent tape
- transparent, 2-inch-wide packaging tape (optional)
- a copy of *Make-It-Yourself Meterstick* (1 per person)
- a copy of the *Make-It-Yourself Meterstick Sheet* on card stock (1 per person)

Using This Activity

To make these measuring devices more durable, you can reinforce the paper metersticks with a long piece of transparent packaging tape.

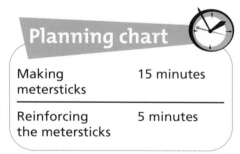

Planning chart

Making metersticks	15 minutes
Reinforcing the metersticks	5 minutes

Make-It-Yourself Meterstick

Follow these instructions to make your own meterstick.

What Do I Need?

◇ scissors

◇ clear tape

◇ *Make-It-Yourself Meterstick Sheet*

◇ clear packaging tape (optional)

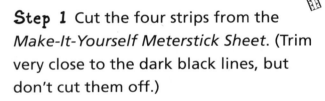

What Do I Do?

Step 1 Cut the four strips from the *Make-It-Yourself Meterstick Sheet*. (Trim very close to the dark black lines, but don't cut them off.)

Step 2 Arrange the strips so that you have a line of numbers that starts with 1 and ends with 100. Tape the strips together in that order. That means you'll tape the strip that ends with the number 25 to the strip that starts with the number 26.

Step 3 After you have finished taping the strips together, your *Make-It-Yourself Meterstick* should go from 1 to 100 in counting order. Each number marks 1 centimeter (cm).

Reinforcing Your Meterstick

Step 1 Have a friend help you do this. Place your *Make-It-Yourself Meterstick* on the table with the numbers facing up.

Step 2 Pull about 30 cm of clear packaging tape away from the roll, but do not cut it.

Step 3 Starting at one end of your meterstick, carefully cover the paper with the tape, a little bit at a time. Smooth out any air bubbles or wrinkles as you go.

Step 4 Pull out more packaging tape as needed until you have covered the entire meterstick with it. Cut the tape when you reach the end of the paper.

Step 5 Fold the end of the tape onto the back side of your meterstick. Do the same with the tape that sticks out along the sides of your meterstick.

You're done!

Make-It-Yourself Meterstick Sheet

TAB

TAB

TAB

The Math Explorer
Published by Key Curriculum Press / © 2003 Exploratorium

Centimeter Ruler

If your group needs centimeter rulers, members can easily make them.

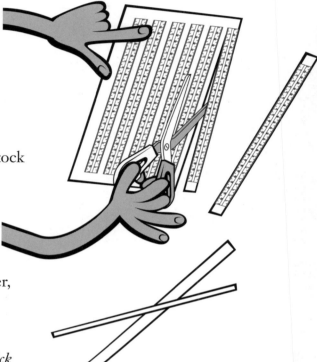

Preparation and Materials

To do this project, you will need:

- a pair of scissors for each person
- copies of the *Centimeter Ruler Sheet* on card stock (1 copy for every 5 rulers)
- transparent packaging tape (optional)

Using This Activity

Have members of your group carefully cut out the rulers. They should trim close to, but not over, the outer borders.

To make the rulers more durable, you can reinforce them with transparent packaging tape. (See the instructions for *Make-It-Yourself Meterstick* on page 200.)

Planning chart

Making rulers	5 minutes
Reinforcing the rulers	5 minutes

Centimeter Ruler Sheet

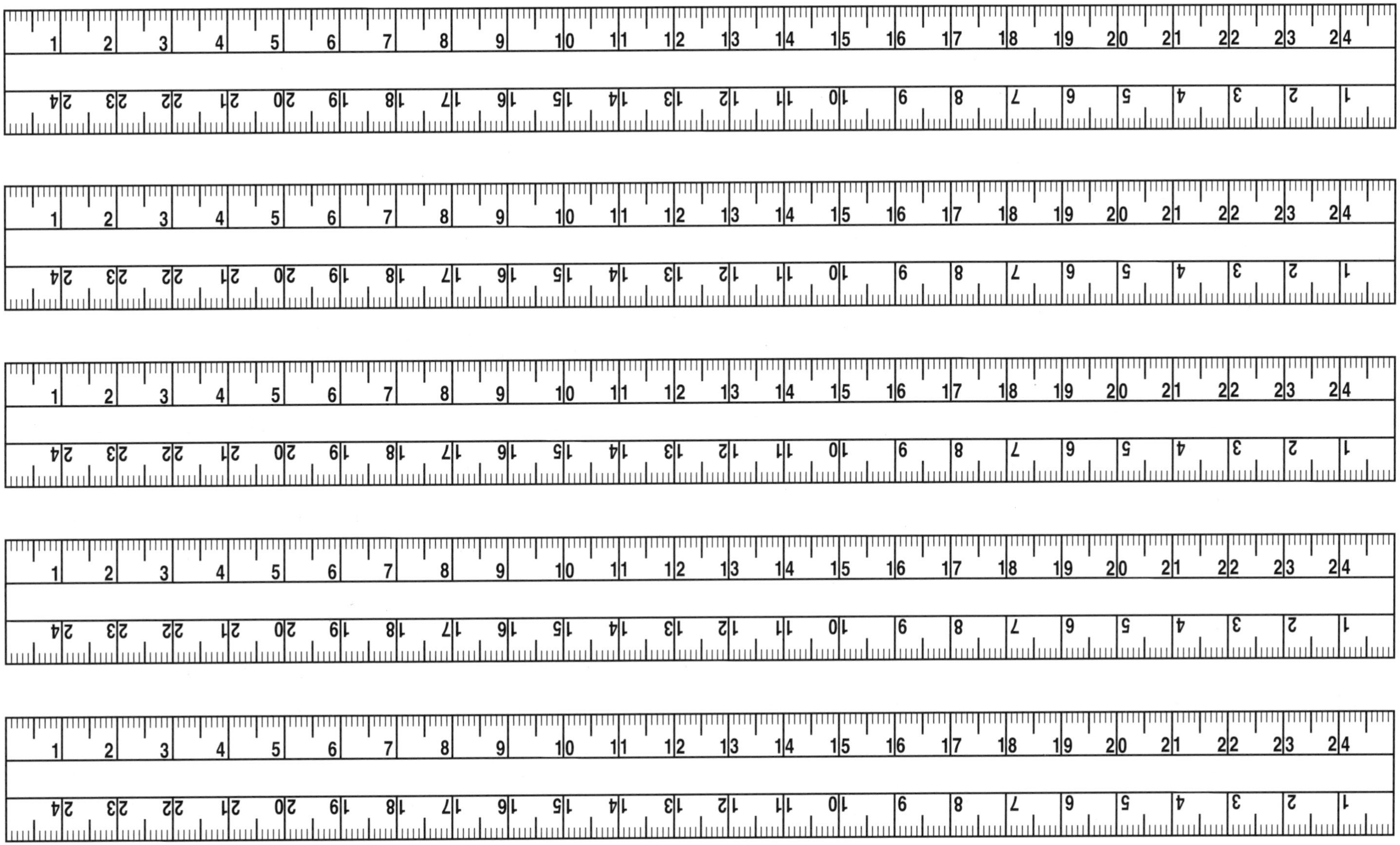

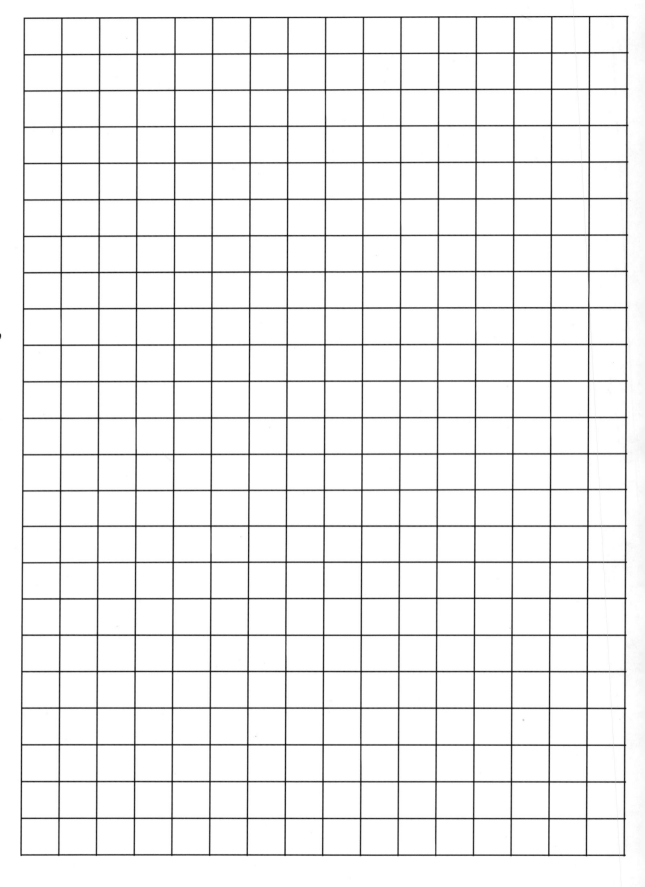

Centimeter Grid Paper

The Math Explorer
Published by Key Curriculum Press / © 2003 Exploratorium

A Note About Metric Conversion

The metric system is the standard of measurement for most people around the world. But many people in the United States are unfamiliar with metric measurements. This table provides numbers you can use to convert metric measurements to inches and miles. For more information about conversions, see page 191.

Unit	Abbreviation	Number of Meters	Approximate Equivalent
kilometer	km	1000 meters	0.62 mile
meter	m	1 meter	39.37 inches
centimeter	cm	0.01 meter	0.39 inch
millimeter	mm	0.001 meter	0.039 inch

NCTM Content and Process Standards

In an effort to improve mathematics education, the National Council of Teachers of Mathematics (NCTM) has established math-content and process standards. These standards are divided into the following content areas: Number and Operations, Algebra, Geometry, Measurement, Data Analysis and Probability; and these process areas: Problem Solving, Reasoning and Proof, Communication, Connections, Representation. In each area, NCTM has established several standards and in the content areas, a set of expectations for each standard.

Activities in *The Math Explorer* address all 10 of the NCTM content and process areas. All 10 areas apply to students in grades six through eight. The following list and the chart on page 209 identify the standards and expectations that are relevant to *The Math Explorer*. In the list, each standard under a given content or process area is identified with a number (1, 2, 3 . . .). The expectations associated with that particular standard are identified with letters (a, b, c). In the chart that follows, we show which of these standards and expectations correspond to each activity. Simply find the activity name in the left column of the chart, and read across to see which content and process areas are covered by that activity.

This text comes from *Principles and Standards for School Mathematics* (NCTM, 2000). You can view the full text of this NCTM publication online at http://standards.nctm.org/.

Published by Key Curriculum Press / © 2003 Exploratorium

Number and Operations

1. Understand numbers, ways of representing numbers, relationships among numbers, and number systems.

   a. Understand and use ratios and proportions to represent quantitative relationships.

   b. Develop an understanding of large numbers and recognize and appropriately use exponential, scientific, and calculator notation.

   c. Use factors, multiples, prime factorization, and relatively prime numbers to solve problems.

2. Understand meanings of operations and how they relate to one another.

   a. Understand the meaning and effects of arithmetic operations with fractions, decimals, and integers.

   b. Use the associative and commutative properties of addition and multiplication and the distributive property of multiplication over addition to simplify computations with integers, fractions, and decimals.

3. Compute fluently and make reasonable estimates.

   a. Select appropriate methods and tools for computing with fractions and decimals from among mental computation, estimation, calculators or computers, and paper and pencil, depending on the situation, and apply the selected methods.

   b. Develop, analyze, and explain methods for solving problems involving proportions, such as scaling and finding equivalent ratios.

Algebra

1. Represent and analyze mathematical situations and structures, using algebraic symbols.

   a. Develop an initial conceptual understanding of different uses of variables.

2. Use mathematical models to represent and understand quantitative relationships.

   a. Model and solve contextualized problems, using various representations such as graphs, tables, and equations.

Geometry

1. Analyze characteristics and properties of two- and three-dimensional geometric shapes and develop mathematical arguments about geometric relationships.

   a. Precisely describe, classify, and understand relationships among types of two- and three-dimensional objects by using their defining properties.

   b. Understand relationships among the angles, side lengths, perimeters, areas, and volumes of similar objects.

2. Specify locations and describe spatial relationships using coordinate geometry and other representational systems.

3. Apply transformations and use symmetry to analyze mathematical situations.

   a. Describe sizes, positions, and orientations of shapes under informal transformations such as flips, turns, slides, and scaling.

4. Use visualization, spatial reasoning, and geometric modeling to solve problems.

   a. Recognize and apply geometric ideas and relationships in areas outside the mathematics classroom, such as art, science, and everyday life.

Measurement

1. Understand measurable attributes of objects and the units, systems, and processes of measurement.

   a. Understand both metric and customary systems of measurement.

b. Understand relationships among units and convert from one unit to another within the same system.

c. Understand, select, and use units of appropriate size and type to measure angles, perimeter, area, surface area, and volume.

2. Apply appropriate techniques, tools, and formulas to determine measurements.

a. Use common benchmarks to select appropriate methods for estimating measurements.

b. Select and apply techniques and tools to accurately find length, area, volume, and angle measures to appropriate levels of precision.

c. Develop strategies to determine the surface area and volume of selected prisms, pyramids, and cylinders.

d. Solve simple problems involving scale factors, using ratio and proportion.

Data Analysis and Probability

1. Formulate questions that can be addressed with data and collect, organize, and display relevant data to answer them.

a. Select, create, and use appropriate graphical representations of data, including histograms, box plots, and scatterplots.

2. Select and use appropriate statistical methods to analyze data.

a. Discuss and understand the correspondence between data sets and their graphical representations, especially histograms, stem-and-leaf plots, box plots, and scatterplots.

3. Understand and apply basic concepts of probability.

a. Use proportionality and a basic understanding of probability to make and test conjectures about the results of experiments and simulations.

Problem Solving

1. Build new mathematical knowledge through problem solving.

2. Solve problems that arise in mathematics and in other contexts.

3. Apply and adapt a variety of appropriate strategies to solve problems.

4. Monitor and reflect on the process of mathematical problem solving.

Reasoning and Proof

1. Make and investigate mathematical conjectures.

Communication

1. Organize and consolidate mathematical thinking through communication.

2. Communicate mathematical thinking coherently and clearly to peers, teachers, and others.

3. Analyze and evaluate the mathematical thinking and strategies of others.

Connections

1. Recognize and apply mathematics in contexts outside of mathematics.

Representation

1. Create and use representations to organize, record, and communicate mathematical ideas.

2. Use representations to model and interpret physical, social, and mathematical phenomena.

The Math Explorer
Published by Key Curriculum Press / © 2003 Exploratorium

NCTM Content and Process Standards Chart

See page 206 for the text of the standards listed on this chart.

Activity	Number and Operations	Algebra	Geometry	Measurement	Data Analysis and Probability	Problem Solving	Reasoning and Proof	Communication	Connections	Representation
1 Boxed In!	2b					2, 3				
2 Oddball		2a			1a, 2a, 3a	1, 2, 3, 4	1			
3 Pig	1a, 3a	2a	3				1			2
4 Madagascar Solitaire			3			1, 2, 3, 4	1			
5 Fantastic Four	1b, 2b							1, 2, 3		
6 Eratosthenes' Sieve	1c, 2a							1		1
7 Hopping Hundred	2a									
8 Tic-Tac-Toe Times	2a									
9 Magic Grid		2a					1			
10 Mind Reader		1				1, 2, 3	1	2	1	1
11 Exponential Folding	1b	2a				1, 2, 3				
12 Colossal Cartoons	3b		3a	2d						
13 Greeting Card Boxes		1a	1a	1a, 2b, 2c					1	
14 Jacob's Ladder						1, 3			1	
15 Paper Engineering			4a	2					1	
16 Incredible Shrinking Shapes	1a, 3b		1a, 1b, 3	2b, 2d					1	
17 Height Sight	3b		1b, 4a	1a-c, 2d		1, 2			1	1, 2
18 Stomp Rocket!	3b		1b, 4a	1a-c, 2d	1	1, 2			1	1, 2
19 Tetrahedral Kites			1b						1	2
20 Flying Things	1a	1a	1b, 2, 4a	2					1	2
21 Stride Ruler				1a, 1b, 2a					1	2
22 Paper Dice					1a					

Glossary

area *(Greeting Card Boxes)*
The number of square units that will cover a surface.

base *(Incredible Shrinking Shapes)*
A base is the number that you multiply in an exponential expression. For example, in the exponential expression 2^3 (which stands for $2 \times 2 \times 2$), 2 is the base. See also *exponent*.

bilateral symmetry *(Madagascar Solitaire)*
Something has bilateral symmetry if it has exactly one line of reflection symmetry.

composite number *(Eratosthenes' Sieve)*
A composite number is a number you find by multiplying two or more numbers.

cubic centimeters (cm³), cubic inches (in.³) *(Incredible Shrinking Shapes, Greeting Card Boxes)*
Cubic centimeters and cubic inches are used to measure volume. A cube with edges one centimeter long has a volume of one cubic centimeter; a cube with edges two inches long has a volume of eight cubic inches.

denominator *(Eratosthenes' Sieve, Hopping Hundred, Stride Ruler)*
The denominator of a fraction is the bottom number. In the fraction $\frac{1}{3}$, for example, the denominator is 3. See also *numerator*.

dimensional analysis *(Stride Ruler)*
Dimensional analysis is a process of treating the dimension (units) as part of an expression and multiplying or dividing one dimension by another to get the dimensions of the answer. You can use it to convert units. For example, $60\text{km} \times \frac{0.62 \text{ miles}}{1 \text{ km}} \approx 37$ miles.

distributive property of multiplication *(Mind Reader)*
The distributive property describes a relationship between multiplication and addition. For example, $(2 + 3) \times 5$ equals $(2 \times 5) + (3 \times 5)$. You can add first, then multiply ($5 \times 5 = 25$) or multiply first, then add ($10 + 15 = 25$).

divides evenly into *(Eratosthenes' Sieve, Hopping Hundred)*
When one number divides evenly into another (the second number is divided by the first), the answer is a whole number and there is no remainder. For example, 4 divides evenly into 12, because 12 divided by 4 is a whole number, 3.

The Math Explorer
Published by Key Curriculum Press / © 2003 Exploratorium

estimate *(Stride Ruler)*

An estimate is a well-informed approximation that's not exactly right, but is close enough for a given purpose.

exponent *(Incredible Shrinking Shapes)*

An exponent tells how many times you multiply the base in an exponential expression. For example, in the exponential number 4^5 (which stands for $4 \times 4 \times 4 \times 4 \times 4$), 5 is the exponent.

exponential expression *(Fantastic Four)*

An exponential expression is an expression that involves exponents. For example, 3^2 is an exponential expression, as is x^3. See also *base* and *exponent*.

factor *(Greeting Card Boxes, Tic-Tac-Toe Times, Hopping Hundred)*

Factors are numbers that you multiply together to produce another number. For example, 3 and 4 are factors of 12, because 3 times 4 is 12.

factorial *(Fantastic Four)*

The factorial of a positive number n is the product of all the positive integers from 1 to n, written as $n!$ For example, 5 factorial is written 5! and is equal to $5 \times 4 \times 3 \times 2 \times 1$, or 120. See also *product*.

fractal *(Paper Engineering)*

A fractal is an infinite pattern formed by an iterative (repeating) process starting with an object such as a mathematical equation or geometric shape, making a change (applying a rule) to that object to produce the first iteration, applying the same rule to the first iteration to produce the second iteration, applying the rule to that iteration, and so on, infinitely many times. The process of creating a fractal is recursive because the result of one iteration becomes the starting point for the next.

iteration, iterative procedure *(Paper Engineering)*

Iteration is repetition. When you solve a math problem or build a pop-up sculpture by iteration, you repeat a process many times.

line of symmetry *(Madagascar Solitaire)*

A line of symmetry is a line that divides something into two mirror images. See also *mirror symmetry* and *reflection symmetry*.

mirror symmetry *(Madagascar Solitaire)*

Something has mirror symmetry if it has a line of symmetry. Mirror symmetry is also called *reflection symmetry*.

multiple *(Hopping Hundred, Eratosthenes' Sieve)*

One number is a multiple of another number if the first number divides evenly into the second number. For example, the multiples of 3 are 3, 6, 9, 12, 15, and so on.

numerator *(Eratosthenes' Sieve, Hopping Hundred, Stride Ruler)*
The numerator of a fraction is its top number. In the fraction $\frac{2}{5}$, for example, the numerator is 2. See also *denominator*.

order of operations *(Fantastic Four)*
The order in which you do the operations in an expression can affect your results. See page 44 for the conventional order in which operations are done.

ordered pair *(Madagascar Solitaire)*
An ordered pair is a pair of numbers you use to locate a point on a coordinate grid. The first number, called the *x*-coordinate, tells how far to the left or right of the origin the point is. The second number, called the *y*-coordinate, tells how far up or down from the origin the point is. The origin is the point with coordinates (0, 0).

place value *(Mind Reader)*
The place value of a digit in a numeral is determined by its position. For example, in the number 213, the 2 is in the hundreds place and represents 200. Likewise, the 1 represents 10, and the 3 represents 3.

prime number *(Eratosthenes' Sieve, Hopping Hundred)*
A prime number is a positive whole number that has only two distinct divisors—itself and one. The first six prime numbers are 2, 3, 5, 7, 11, 13.

probability *(Pig)*
The probability of an event is a number between 0 and 1 that expresses the likelihood that that event will happen. Probabilities can be expressed as fractions. For example, if you flip a fair coin, the probability of getting heads is $\frac{1}{2}$. If you roll a six-sided die, the probability of rolling a 2 is $\frac{1}{6}$.

product *(Tic-Tac-Toe Times)*
A product is the result of multiplying two or more numbers.

proportion *(Incredible Shrinking Shapes)*
A proportion expresses two ratios that are equal. For example, the ratios $\frac{3}{6}$ and $\frac{4}{8}$ form the proportion $\frac{3}{6} = \frac{4}{8}$.

ratio *(Pig, Incredible Shrinking Shapes, Flying Things)*
A ratio expresses a comparison of two numbers. For example, if there is one apple and three oranges, the ratio of apples to oranges is 1 to 3. This ratio can be written as 1 to 3, or 1:3, or $\frac{1}{3}$.

The Math Explorer
Published by Key Curriculum Press / © 2003 Exploratorium

recursive *(Paper Engineering)*

A recursive process is a repeated process where the result of one iteration is used to determine the next iteration.

reflection symmetry *(Madagascar Solitaire)*

Something has reflection symmetry if it has a line of reflection. Reflection symmetry is also called *mirror symmetry*. See also *line of symmetry*.

rotational symmetry *(Madagascar Solitaire)*

Something has rotational symmetry if you can rotate it about its center less than 360° and have it look the same as it did in the original orientation. (If you rotate something 360°, it ends up back in its original position—and of course it looks the same!)

rounding *(Flying Things)*

You use rounding to reduce the accuracy of a calculation. Round an answer when you've included measurements in a calculation, so that the results don't appear to be more precise than the original measurements were. For example, if you measured to the nearest tenth of a centimeter (one place to the right of the decimal point), you should also round the results of your calculations to the nearest tenth of a centimeter. *Rounding* is explained in detail on page 185.

similar figures *(Height Sight)*

Two figures are similar if they are exactly the same shape, but not necessarily the same size. Any two squares are similar. Two rectangles are similar only if the ratios of the sides are equal.

square centimeters (cm²), square inches (in.²) *(Incredible Shrinking Shapes, Greeting Card Boxes)*

Square centimeters and square inches are used to measure area. A square with sides 1 centimeter long fills an area of 1 square centimeter; a square with sides 2 inches long fills an area of 4 square inches.

square root *(Fantastic Four)*

Finding or "taking" a square root is the opposite of squaring a number. To square a number, multiply that number by itself (2^2 is 2×2, or 4). To take the square root, find the number that, when multiplied by itself, gives the original number. For example, the square root of 4 is 2, because 2 squared (2×2) is 4. A square root is designated by the symbol $\sqrt{}$. For example, $\sqrt{81} = 9$.

volume *(Greeting Card Boxes)*

The amount of space (number of cubic units) an object occupies.

Credits and Acknowledgements

The Exploratorium would like to thank the National Science Foundation for the grant that made this book possible

The Exploratorium team would also like to thank the members of our board of advisors: Joe Buhler, Ron Lancaster, Robert V. Lange, Alec L. Lee Jr., Hugo Rossi, Jim Sandefur, Justine Underhill, Ellen Wahl, and Faedra Lazar Weiss.

We sought advice from youth group leaders, program administrators, and curriculum developers. We appreciate the assistance of Mary Ager of Girls Inc. of Alameda County; Vivian Altmann of the Exploratorium's Children's Educational Outreach Program; Sue Eldridge and Sam Piha of the Community Network for Youth Development; Jean Fahey of the Girl Scouts; Don Garcia of AIM High; Jumoke Hinton of Girls After School Academy; Eileen Murphy of Back on Track; Michael Funk of Sunset Beacon Neighborhood Center; Tara Wilson and Meenha Lee of Boys and Girls Clubs of San Francisco; and Fran Chamberlain and Laurel Robertson from the Developmental Studies Center.

Activities for this book came from a variety of sources. The game we call *Hopping Hundred* is based on a game developed by Rob Porteous, currently a math teacher at St. George's School for Girls in Edinburgh, Scotland. A dedicated group of teachers, parents, and youth development leaders helped find and develop a variety of activities: Marisa Alfieri, Susan Audap, Paul Barron, Ken Beckman, Eugene Berg, Paula Bosque, Rilla Chaney, Carolyn Harpster, Jumoke Hinton, Trish Mihalek, Patricia Tucker, Justine Underhill, Eric Watterud, and Clemency Wings.

A variety of youth groups tested activities for us, providing feedback and welcome suggestions for improvements: African American Family Education and Cultural Center, Boys and Girls Clubs of San Francisco, Community Housing Opportunities Corporation, Excelsior Beacon Center, Gateway/New Perspectives, Girls Adolescent Program of the Chinatown Youth Center, Girls Inc. of Alameda County, Girls Inc. of Memphis, Girls Inc. of Omaha, Presidio YMCA, St. John's Tutoring Center, and Sunset Beacon Neighborhood Center. Jeanne D'Arcy of Aptos Middle School and Jennifer Bloomer of Moss Landing Middle School tested these materials in their classrooms and in after-school programs.

In addition, the following folks tested materials at home:

Kelcie Angstadt	Amy Bridges	Kristin Delaney	Addison Gregory
Megan Angstadt	Myles Bugbee	Trey Dillow	Bradley Griffin
Michael Antonio	Craig Burkhert	Lyell Doll	Tiffany Griffin
Robin Armstrong	Elia Caffey	Sam Ellis	Alex Haas
Simon Armstrong	Joshua Campbell	Rael Enteen	Chase Haegele
the Arnold Family	Alexa Carey	Shaina Epstein	Elaine Haegele
Sally Baldwin	Jesse Carlson	Alex Farmer	Matthew Hale
Emma Bergmann	Jason Chew	Aliza Fegan	Kevin Hall
Laura Bergmann	Patrick Chew	Peter Florin	Jeffrey Hargrove
Ryan Bleezarde	Ahn Min Choi	Chris Fong	Jeremy Higgins
Devin Bletterman	Luc Cote	Satchel Friedman	Kyle Higgins
Tom Bliska	Max Countryman	Will Friedman	Jessica Hinze
Mike Bonner	Dimitri de Kouchkovsky	Rebekah Gleber	Ellen Huet
Max Braunstein	Dustin Delaney	Carly Greenberg	Lucie Huet

(continued)

Kelley Hunter's Windsor
  Middle School class
Austin Issler
Chris Johnston
Daniel Johnston
Matt Johnston
Drew Jones
Zachary Jones
Leah Karlins
Naimi Karp
Jenna Keith
Alex Kohan
Mara Kubrin
Nina Lagpacan
Merlin Larson
Andrew Lee
Catherine Lee
Christina Lee
Matthew Lee
Everett Lee-Wuollet
Christian Lemus
Zachary Lerman
Joanna Lichtin
Kyia Lively
John Luikart
Kierstyn Lyts
Greg Macaulay
Kirsten Macaulay
Karina Maravelias
Stephanie Marshall

Camila Martin
Richard Matthews
Lucia Mattox
Ava McLeod
Caroline McMillan
Katherine McMillan
Avi Miller
Lyle Mills
Henry Milner
Jenny Miner
Alina Monaghan
Jeremy Moody
Derek Morgan
Nick Morrill
Tyler Moskovitz
Aaron Motley
Alexis Motley
Colin Nicol
Hans Nielson
Kelly Nielson
Troy Oliver
Alexander O'Sullivan
Sergio Pardo
Mary Paulson
Patrick Peterson Smith
Sara Phetteplace
Brynna Popka
Robin Portillo
Alexandra Presher
Nicole Presher

Jesse Purvey
Kristina Richardson
Noah Rose
Amy Rothermund
Carissa Rowley
Matthew Rowley
Emily Rulla
Matt Schroeder
Nikos Schwelm
Spencer Scott
Stephanie Scott
Jen Sedor
Julia Seeholzer
Matthew Shaw
Tim Shaw
Tim M. Shaw
Marcie Sheperd
Marcia Sherry
Bryant Shreeve
Steven Shreeve
Eric SooHoo
William Spencer
Jonathan Spivack
Spencer Stamats
Peter Stark
Lloyd Steel
Alexandra Stockton
Christina Stockton
Stephanie Stockton
Aaron Strick

Derek Strick
Andrew Sutherland
Jessica Sweeny
Hawnlay Swen
Laine Tennyson
John Thuemler
Emily Tow
Rebecca Varon
Bianca Velez
Andrew Vendl
Sean Vincent
Jessica Vischansky
Stephanie Vischansky
Brandon Waters
Ryan Waters
Simon Watsky
Alexa White
Danny Wilkes
Richard Wille
Joseph Winn
Khylen Wood
Dave Woodbury
Kristina Woodbury
Mark Woodbury
Rebecca Woodbury
Wyatt Yurth
Zachary Zemarkowitz

About the Exploratorium

The Exploratorium is a hands-on museum of science, art, and human perception that is dedicated to discovery. Founded in San Francisco in 1969 by the noted physicist and educator Frank Oppenheimer, the museum has grown over the years to become an internationally acclaimed science center. Its hundreds of interactive exhibits stimulate learning and richly illustrate scientific concepts and natural phenomena.

The Exploratorium's Center for Teaching and Learning offers a teacher-centered, learn-by-doing approach to science instruction. Working with teachers and after-school groups, Exploratorium staff members use museum exhibits and classroom activities to explore concepts in science and math and to help people learn by asking questions and experimenting.

Through books like this one, the museum makes our favorite activities available to people who can't visit the museum. You can learn more about our publications and find other valuable math and science learning resources on the Exploratorium Web site at www.exploratorium.edu.

About the Authors

Pat Murphy has written a number of science activity books for kids, including *The Science Explorer*, *The Science Explorer Out and About*, and *The Brain Explorer*. She has been a writer and an editor at the Exploratorium since 1983.

Lori Lambertson is an avid surfer, an artist, and a former bicycle racer who has been teaching middle school math and science for more than a decade. She currently works in the Exploratorium's Teacher Institute as a staff teacher and as coordinator of the Exploratorium's New Teacher Program. She says she has the world's best job.

Pearl Tesler is a science writer with coordinates in San Francisco. She clearly recalls crying in frustration while (not) learning negative numbers in sixth grade, but has since gone on to integrate, diagonalize, and normalize with positive pleasure.